A FIELD GUIDE TO FOSSILS & GEOLOGY of the Isle of Wight

MEGAN JACOBS

A Field Guide to Fossils & Geology of the Isle of Wight
by Megan Jacobs

First published in 2026 by Medina Publishing Ltd.
50 High Street, Cowes, Isle of Wight, PO31 7RR, England

www.medinapublishing.com

Medina Publishing

Printed and bound in Lithuania by Offset Printing House KOPA

Paperback edition
ISBN-13 : 978-1916866218

Also published as a limited edition hardback
ISBN-13 : 978-1916866140

CIP data: a catalogue record for this book is available
at the British Library.

Typeset in EB Garamond 11/15.
Captions in Roboto.

A FIELD GUIDE TO FOSSILS & GEOLOGY of the Isle of Wight

MEGAN JACOBS

Medina Publishing

Contents

Michael J. Benton

Foreword

The Isle of Wight is renowned for its diverse geology, which spans primarily Cretaceous and Cenozoic rocks. Over the years, there have been numerous books and articles devoted to the geology and palaeontology of the island as well as field guides to the whole island, or about specific themes such as the Wealden, the Palaeogene, or the Pleistocene. Much of the fame of the island as a geological destination in recent years has come from its dinosaurs, of which some 15 genera have been named, notably *Iguanodon*, *Hypsilophodon*, and *Polacanthus* in Victorian times, and a further nine this century.

In this book, Megan Jacobs provides an eloquent account of the history of geological and palaeontological study of the island, and the fact that many great names in the field took an interest. Whether Queen Victoria, some-time denizen of Osborne House on the island, ever explored the dinosaur sites is not recorded.

The local and wider interest in the geology and fossils of the island is reflected in the activities of the Isle of Wight Natural History Society and other bodies, which have been responsible in various ways for the documentation of discoveries and their display in a series of museums. The Museum of Isle of Wight Geology in Sandown opened in 1911, and this was succeeded by the Dinosaur Isle Museum in 2001, where many of the key fossils, including type specimens of the newly named dinosaurs, are to be seen.

Megan Jacobs provides detailed context to other geological features of the Isle such as springs, chines, and sea stacks, all of which have been attractions to tourists as well as important indicators of the underlying geology. The geology is described, with enough basic background information to allow a non-geologist to understand the details of the major Isle of Wight Monocline, an east-west structure that places the younger Cenozoic rocks at the surface in the northern half of the island, and then steps up to expose the Cretaceous, mainly Wealden and Chalk in the southern half.

The sedimentology and stratigraphy of the Cretaceous and Cenozoic succession is established in considerable detail, as are the environments of deposition and the indications of climate change recorded through the past 130 million years on the island. These geological features are engagingly described and vividly illustrated by photographs. Late chapters describe the basics of fossil collecting, what to do with your fossils after they have been collected, and how to identify and interpret them.

There is enough in this book for the novice geologist or palaeontologist of any age. The final chapters help the reader to identify and interpret the fossils and provide indications of further information. Over the past two centuries, thousands of budding geologists and palaeontologists have taken their first scientific steps on the Isle of Wight, and this book should help the next generation to appreciate the wonders of what they might see and collect, and how these can provide a deeper understanding of the wonders of the history of the Earth and the history of Life.

Michael J. Benton
Professor of Vertebrate Palaeontology
University of Bristol
June 2025

1. HISTORY

Figure 1.1A
Robert Hooke (1635-1703)

1.1 Isle of Wight Palaeontology

The Isle of Wight's palaeontological fame began in the 17th Century with a young Robert Hooke (1635-1703) (*Fig. 1.1A*) who was born in Freshwater. At age 13, Hooke moved to London to pursue an education, resulting in attending Oxford University, then employment at the Royal Society as Curator of Experiments as well as Professor of Geometry at Gresham College. Hooke worked very closely with Isaac Newton. Hooke went on to make his own microscope, which later resulted in comparing the microstructure of fossil wood and modern wood. This led him to conclude that fossils are the remains of once-living things that had been soaked in petrifying water laden with minerals (now known as diagenesis). He published his observations and conclusions of his microscope work in a book called *Micrographia*. One of the most important scientific works ever written.

As a child on the Isle of Wight, Hooke had a keen interest in geology and collected fossils. Spending many hours exploring and observing the cliffs at Freshwater Bay and the section from Alum Bay to Colwell. It's clear his local beaches and the fossils he found influenced his ideas on prehistory and geology in later life.

It was already accepted at that time that certain fossils found close to a sea or lake, and resembling living creatures were the remains of long-dead organisms. However, there was a whole body of fossils that were much harder to explain: fossils that bore no resemblance to anything living today and, more especially, fossils of marine animals that were now found far inland or high in the hills and mountains. It was these anomalous fossils that Hooke sought an explanation for.

Between 1667 and 1669, he delivered a series of lectures on earth science entitled "*Lectures and Discourses of Earthquakes and Subterraneous Eruptions. Explicating the Cause of the Rugged and Uneven Face of the Earth; And What Reasons may be given for the frequent findings of Shells and other Sea and land Petrified Substances, scattered over the whole Terrestrial Superficies.*" It is an exceptionally long title, but then again, he was discussing a very extensive and far-reaching topic.

During these talks, Hooke proposed that the Earth's surface had been changed due to earthquakes and volcanoes, providing an explanation as to why fossils can be found in places where the animals clearly didn't live (e.g., seashells at the top of a mountain). During these lectures, Hooke used the fossils he'd found in the cliffs and hills around Alum Bay to support his arguments for these geological changes.

A more extensive description of the island's geology and some fossil specimens were published by Sir Henry Englefield in 1816, based on the geological work of Thomas Webster. However, a few years later, palaeontological focus on the island had moved from geological based studies to the larger vertebrate fossils. James Hay is said to have first reported dinosaur bones on the island. He had acquired a collection of vertebrae and limb material from near the parish of Motteston (Mottistone), and stated they were from a 'stupendous animal, supposed to be the Mammoth or Mastodon". Mr. Hay went on to amass a huge collection of fossils, leading him to conclude that these animals once inhabited southern England at a 'very remote period'.

In 1829, William Buckland (*Fig 1.1B*) described some *Iguanodon* remains found on the beach at Yaverland near Sandown. He also noted *Iguanodon* bones found by Mr. James Vine at Brook. James Vine, a London born businessman, and Treasurer of the Geological Society of London, moved to the island between 1812 - 1825. He built Puckaster Cottage, on the Undercliff near Niton, allowing him to fossil collect along the island's southwest coastline. It's said Vine's attention were brought to the dinosaur bones by a local collecting them to be put into a grotto, noting the pyrite and calcite infilling of the bones. If only dinosaur bones were common enough now to be used as a building stone!

Figure 1.1B
William Buckland (1784-1856)

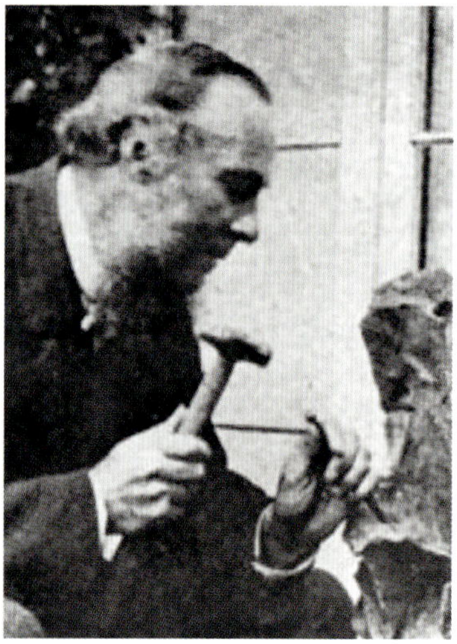

Figure 1.1C
Reverend William Fox (1813-1881)

By the early 19th Century, palaeontological focus on the island had moved onto the larger vertebrate fossils. This began in 1829 when William Buckland (*Fig. 1.1B*) described some *Iguanodon* remains found on the beach at Yaverland near Sandown. However, the most famous island fossil collector at the time was the Reverend William Fox (1813-1881) (*Fig. 1.1C*); the curate at Brighstone church. Fox spent much of his time walking the southwest coast of the island, becoming a prolific collector of vertebrates from the varicoloured clays and sandstones that form the cliffs and wave-cut platforms on this beautiful part of the island's coast.

Fox provided many fossils for Sir Richard Owen (1804-1892) (*Fig. 1.1D*) at the British Museum of Natural History and Gideon Mantell (1790-1852) of Lewes, Sussex (*Fig. 1.1E*), whom both described his discoveries in the scientific literature of the day. Reverand William Fox is credited with discovering the dinosaurs *Aristosuchus, Calamospondylus, Hypsilophodon* and *Polacanthus* although he never described them, leaving that job to the experts. Fox's love of collecting remained with him until he died, famously saying 'I cannot leave this place while I have any money left to live on,' concluding 'I take deep pleasure in hunting for old dragons.' Fox kept his word and spent the rest of his life in Brighstone village hunting old dragons. His body now rests in the graveyard of St. Mary's Church in the heart of the village. During his time on the island, Fox accumulated a diverse collection of over 500 specimens, which were acquired by the Natural History Museum, London after his death in 1882. His collection remains housed in the museum and has been studied on numerous occasions since, still providing an important reference for modern studies on dinosaurs.

Figure 1.1D

Sir Richard Owen (1804-1892)

Figure 1.1E

Gideon Mantell (1790-1852)

Around this period, in 1858, Charles Darwin holidayed on the island with his family. It is during this holiday, based in Sandown, it's said he began writing his most famous work – *On the Origin of Species*. A work that was nothing short of revolutionary, changing the way we understand the natural world and challenged popular notions of divine creation.

In 1830, Bishop Samuel Wilberforce, otherwise known as 'Soapy Sam', was transferred to St. Mary's church in Brighstone (*Fig. 1.1F*). Wilberforce, a devoted man of God, famously took part in the debate concerning evolution in 1860 at the University of Oxford. The debate saw Wilberforce and Thomas Henry Huxley (*Fig. 1.1G*) (often referred to as Darwin's bulldog), discuss Darwin's controversial new book. Wilberforce remained strongly opposed to the concept of evolution, especially the idea humans were descended from apes, asking Huxley 'whether it was through his grandfather or his grandmother that he claimed his descent from a monkey'. Huxley's reply (paraphrased), "I would rather be descended from an ape than an ignoramus". Sadly, Darwin was too sick to hear the heated debate.

Figure 1.1F

Bishop Samuel Wilberforce (1805-1873)

Oddly enough, Huxley was no stranger to the Isle of Wight's fossils, and in 1869 he described a new species of a small herbivorous dinosaur. The first specimen was recovered by workers in 1849, in the Cowleaze Chine area, and consisted of a large block with a semi-articulated skeleton. This is now called the Mantell-Bowerbank block, due to the block being split in two, and given to separate people, James Scott Bowerbank, and Gideon Mantell, until being reunited a few years later. In 1855, Richard Owen published a short note on the block, describing it as a juvenile *Iguanodon*.

Figure 1.1G

Thomas Henry Huxley (1825 – 1895)

In 1867, Huxley compared the bones more closely to those of known *Iguanodon*, noting many differences. In 1868, at an exhibition in Norwich, he saw a skull found by William Fox in the same area as the Mantell-Bowerbank block. The skull was notably very different to *Iguanodon*, and upon receiving permission to study the skull from Fox, he began work on the new dinosaur. In 1869, less than a year later, he announced and published in print the new species *Hypsilophodon foxii*. Both the Mantell-Bowerbank block and the skull remain housed at the Natural History Museum, London.

Figure 1.1H
Samuel Beckles (1814-1890)

At Hanover Point on the southwest side of the island, foot casts, made primarily by herbivorous dinosaurs such as *Iguanodon*, were first documented by Gideon Mantell in 1847, who described them as 'concretions' (*Fig. 1.2*). Mantell was reluctant to accept dinosaurian trackmakers, which is puzzling, given his extensive work on *Iguanodon* bones, including the largest toe bone of an *Iguanodon* found to that date. This was found on a garden path in Brixton (Brighstone), so the size of the footprints would not be a factor.

Figure 1.1I
Edward Forbes (1815-1854)

His writings also reveal his frustration with local practices, noting with disapproval that villagers collected fossils for use in paving paths and gardens, though he admitted being too occupied professionally to "remedy the evil mentioned in the text" (i.e., go round removing all the fossils from everyone's gardens in Brighstone). It wasn't until 1853, a year after Mantell's death, that the connection was made between the

Figure 1.2
Iguanodontid foot casts at Hanover Point.

3 toed casts and the skeleton of a dinosaur's foot. Samuel Beckles (*Fig. 1.1H*), a lawyer-turned fossil collector, showed Sir Richard Owen an almost complete hindlimb skeleton from a juvenile *Iguanodon* from the Isle of Wight. The ever-insightful Owen matched the foot to the footprint, possibly marking the birth of vertebrate palaeoichnology (the study of fossilised tracks, traces and burrows) as a discipline in Britain.

Whilst some of the greatest minds in palaeontology at the time focused on the dinosaurs of the southwest coast of the island, Edward Forbes (1815-1855) (*Fig. 1.1I*), a remarkable palaeontologist and natural historian, directed his attention to the north coast of the island and its younger strata. After visiting the island on numerous occasions between 1848 and 1853, he published two monographs documenting the stratigraphy and fossils of the islands north coast, the first in 1853 and a second in 1856. Forbes's work was pioneering in our understanding of the north coast geology and remains to this day as one of the greatest works on the Isle of Wight's geology in the 19th century.

The first half of the 1900s saw a lull in palaeontological discoveries on the island after the scientific boom of the 1800s.

A study on *Iguanodon* and the unusual blunt-snouted pterosaur *Istiodactylus* by Reginald Hooley (1886-1923) is the only work of the early 1900s of any notable importance. Hooley, a wine merchant from Southampton, had a keen interest in palaeontology. He discovered an almost complete skeleton of a small iguanodontid in the Vectis Formation at Atherfield, naming it *Iguanodon atherfieldensis* in 1925. More recently (2012) the specimen was placed into the new genus *Mantellisaurus*. This skeleton is now displayed in the Natural History Museum, London, in the Hintze Hall.

In the 1960s, the study of Isle of Wight dinosaurs came bouncing back, at least in the scientific world, due to research on the small ornithopod dinosaur *Hypsilophodon* by Peter Galton and a study of the ankylosaur *Polacanthus* by William Blows. Also at this time was the discovery of an unusual dinosaur skull bone at Yaverland, near Sandown. This unique specimen was regarded as a new genus and species and was christened *Yaverlandia bitholus* in 1971 by Peter Galton.

Such was the pace of discovery of dinosaurs on the island that by the 1990s the island became known as 'Dinosaur Isle'. In 1992, a large skeleton of a sauropod dinosaur was found projecting from the cliffs at Barns High, opposite what is now known as the Dinosaur Farm.

With new dinosaur skeletons being found and hitting the press, combined with the release of the box office phenomenon Jurassic Park in 1993, dinomania hit the island. Tourists and palaeontologists started heading to the island to search the fossil-rich beaches and examine museum specimens.

In 1996 the first new theropod dinosaur was named on the island for over a century – *Neovenator*. Since then, 9 new dinosaurs have been discovered on the island: *Eotyrannus* (2001), *Vectaerovenator* (2020), *Brighstoneus* (2021), *Vectiraptor* (2021), *Riparovenator* (2021), *Ceratosuchops* (2021), *Vectipelta* (2023) and *Vectidromeus* (2023), *Comptonatus* (2024) and *Istiorachis* (2025). There have also been new species of pterosaurs, *Caulkicephalus* (2005) and *Wightia* (2020), and a crocodile with an almost unpronounceable name, *Kompiodontosuchus*. Other studies include turtles, mammals, amphibians, fish, molluscs, insects, and many more. Amber has also been discovered within the

Wessex Formation on the island, some containing inclusions such as flies and a rare spider.

Almost all the spectacular palaeontological discoveries are thanks to the hard work and diligence of the island's fossil collectors who, over the last 40 years, have dedicated their time and energy to rescuing fossils from being destroyed by the sea and kindly donating them to museums for scientific research.

1.2 Isle of Wight Geological Museum

Of particular importance for the Isle of Wight palaeontology has been the presence of a museum since the early 1800s. The first collections were built by the Isle of Wight Natural History Society, which became the Isle of Wight Philosophical Society. From 1818, the collections were housed in the now Isle of Wight County Club in Newport. Not much is known about the collections at the time, however, it was noted that Reverend Edmund Kell was a curator for a period.

In 1852, Dr. Ernest Wilkins, a philanthropist and physician, took over the care of the collection. He raised funds for the collection to be properly displayed and developed, and moved to Newport Guildhall in 1853. Dr Wilkins even produced a catalogue, providing an idea of the specimens housed in the collection at the time, some of which are still in the collections today. Dr. Wilkins remained curator until his death in 1881.

Following Dr. Wilkins's death, the future of the collection remained uncertain. In the latter part of the 19th century, the collection, along with Dr. Wilkins' private collections, was purchased by the Newport Young Men's Literary Society. Mr John Wood acted as curator of the collections and began cataloguing all the specimens until his death in 1887. However, by 1911, the society decided caring for the collection was too expensive and wanted to use the space for other purposes. The collection was sold to a local businessman Frank Morey, who wanted to keep the collections on the island, for the tidy sum of £100 (about £15,000 in today's money).

Frank Morey donated the geological specimens to Sandown Town Council and the rest of the specimens to Carisbrooke Castle

Museum. Situated above the Sandown town library, the new museum of Isle of Wight Geology opened its doors to the general public in 1913, with a grand total of 593 specimens in the collection.

The first curator of the Museum of Isle of Wight Geology was Mr Hubert Poole, who maintained the collection until this role was taken over by Reverend J.C. Hughes. Hughes continued to manage the collection until his death in 1929. In 1924, Frank Morey hired a second curator, James Frederick Jackson. Jackson, a keen fossil collector, invested much time in expanding the collection personally, donating 2073 specimens from all over the island. He also produced the first complete catalogue of the museum's specimens, organised by stratigraphic horizons (where they were found) and then taxonomically.

Frank Morey died in 1925, and Jackson continued in his position until the death of Frank's sister, Catherine Morey in 1943, who had continued to finance the curatorial position. This left the museum without a curator until 1954 when Alan Grapes took up the position of honorary curator.

It wasn't long until Alan Grapes, with the aid of Mr Jackson's catalogue, realised that many fossils had disappeared, likely stolen from the museum. For about 10 years, with no permanent staff, unsecured cases and fossils often left beneath cabinets, it was easy pickings for 'souvenir hunters'. In total, the museum lost 20% of its collection, a staggering total of 976 specimens. It was thought that academics or collectors were the likely culprits, due to many of the specimens lost being of obscure scientific value, only being of interest to a professional palaeontologist, many were rare, or undescribed taxa., whilst trophy pieces such as dinosaur claws and teeth were left behind. More worrying still, was that no library staff or members of the public noticed any of the thefts, including large *Tropaeum* ammonites which can reach up to 50 cm in diameter, and weigh up to 250lbs (113.4kg).

Alan Grapes retired from his curatorial position due to ill health in 1966. During his time at Sandown, he secured the collections and published a guidebook and several scientific papers in the local journal. He was also known for his love for outreach, engaging and encouraging school children and the public to collect fossils, and giving museum tours to visitors. He also worked with inmates of Camp Hill prison, giving talks on geology and encouraging the inmates to collect fossils, which resulted in 30 fossils being donated to the museum by them.

After Alan's retirement, the museum was run by a series of at least 3 curators: Albert Baker, S. Richardson and N. Preece. It has been reported that many thefts occurred before Richardson took over, following the same pattern as before with only scientifically important specimens going missing. This time cabinets were prised open, and their contents removed.

In 1976, Dr. Allan Insole, a qualified geologist, assumed the role of curator. He was joined shortly after by Steve Hutt in 1978. Steve is a self-taught palaeontologist and fossil collector, noted for finding the Barns High Sauropod in 1992, among many other discoveries.

During the 1980's it became clear that the museum, still located on the upper floor of the town library, was not fit for purpose. It was cramped and the displays were becoming tired and dated.

Figure 1.3

The Dinosaur Isle Museum, Sandown

Visitors had come to expect full-size dinosaur skeletons and interactive displays, seen in the larger museums around the country. In 1987, the museum underwent a redesign, funded by a well-known journalist and TV presenter Magnus Magnusson. This revamp included a full-size wall mount of a large *Iguanodon* discovered and excavated by Hutt, which is now on display in the new Dinosaur Isle Museum.

Between the 1980s and 1990s, the number of visitors and the collection continued to grow in number, but also in national importance, with the donation of several near complete dinosaur skeletons, including holotype material. However, over time, the ceilings of the outdated facility in the library started to collapse under the weight of the collection and the increasing visitor numbers. In addition, space was rapidly running out as new discoveries continued to arrive.

In 1998, the museum received £1.3 million in funds from the Heritage Lottery Fund, along with grants from the Isle of Wight Council and local businesses, totalling £2.7 million, to build an all-purpose museum to house the ever-expanding collection of fossils. By 2001, a new and dedicated palaeontological museum on Sandown seafront was complete, and Dinosaur Isle Museum opened its doors to the public. Today the venue is a major attraction for tourists visiting the island and for scientists researching the island's rich palaeontological heritage (*Fig. 1.3*). There is a charity called 'Friends of Dinosaur Isle (or FODI) that has yearly membership and runs events throughout the year to raise public awareness of the island's palaeontology. All funds raised are used to directly support the museum with acquiring and preserving the fossils in their care.

Cultural history of geology on the Isle of Wight

Despite a dominating academic perspective on the island's geology and fossils in the last 200 years, the geology of the island has shaped our landscape, livelihoods and culture. In the next sections, we explore geological phenomena which have shaped our island's history.

1.3 Chalybeate Springs

Chalybeate springs or ferruginous waters are mineral spring waters containing salts of iron. These springs were reputed to provide health-giving waters, with many physicians in the 17th century promoting the numerous benefits of drinking mineral waters, particularly water from the chalybeate springs.

A physician talking about a chalybeate spring in Tunbridge Wells said chalybeate waters could cure:

> *"The colic, the melancholy, and the vapours; it made the lean fat, the fat lean; it killed flat worms in the belly, loosened the clammy humours of the body, and dried the over-moist brain."*

During the 17th and 18th centuries, two chalybeate springs were discovered on the island: the Sandrock Spring and the Shanklin Spa. Both springs became popular tourist spots, bringing people to the island to take advantage of the supposed benefits of the iron-rich waters.

Sandrock Spring, Niton

Dr Thomas Laurence Waterworth, a physician from Newport began exploring the Niton and Blackgang area in 1807. He soon

Figure 1.4A
View of the Sandrock chalybeate spring and grotto taken from the Brannon (1874).

discovered a spring, with waters deriving from the Sandrock Formation of the Lower Greensands Group. He was granted a lease by the landowner to build a grotto at the spring so that visitors could take sample water (*Fig. 1.4A*). In 1811, he described the spring's discovery and its medicinal properties, touting the virtues of the healing waters. Visitors flocked to the area to sample the healing water and soon the Royal Sandrock Hotel was built near Niton Undercliff. However, by 1860, the demand for the healing water declined, leading to the closure of the spring dispensary. In 1958 the grotto was lost to a landslip, and the spring was lost in 1978 to a further landslip.

Figure 1.4B

Chalybeate springs. Photo showing The Grotto, Royal Spa Hotel, Shanklin

Shanklin Spa

The spring at Shanklin was initially discovered in the 17th Century but wasn't commercially exploited until 1874. This may have been in part due to the difficult access to the spring, as it was located at the base of the cliffs, some 60 metres below the level of the town, just north of The Esplanade today.

Mr Archibald Hinton opened the Royal Spa Hotel in the summer of 1874, taking full advantage of the spring below (*Fig. 1.4B*). He built a reservoir at the rear of his hotel for the spring water, allowing for the water to be piped into the hotel and onto the esplanade, allowing the public to drink it without charge. It is said the spring was discharging at a constant 2000 gallons per day, permitting for a small pump house to be erected to bring a constant supply of water to the hotel for both bathing and drinking. At this point, there was a charge of 1 penny (£2 in today's money) for unlimited consumption of the water for the hotel visitors and the general public.

Both springs on the island yielded water with high concentrations of iron and aluminium (*Fig. 1.4C*). The concentrations in the Sandrock Spring were two

Figure 1.4C

Photo showing the iron-rich chalybeate spring waters flowing in the streets of Shanklin today.

Photo credit to Ray Girvan.

orders of magnitude higher than that of the Shanklin Spa, making the maximum intake recommended just 1 pint a day. This likely contributed to the Sandrock Springs's demise, as it would have been almost undrinkable.

The difference in hydrochemistry of the two springs is due to the geology of the source rock. The waters of the Sandrock Spring are derived from a clean quartz sand, devoid of any reactive silicate or carbonate minerals. However, the water-rock interactions involved the oxidisation of pyrite (an iron sulphide), resulting in the spring waters having a pH of <4.5 or less. Meaning it would have been very acidic which allowed the waters to hold higher concentrations of iron and aluminium.

The Shanklin Spa waters, however, were derived from rocks with a higher abundance of carbonates, leading to a neutral pH of 6 – 7, and much lower levels of iron and aluminium. Overall, this high pH made the water somewhat safer to consume.

However, by the 1900s interest in the spring water diminished, and the spas and springs disappeared from local guidebooks. The medical field showed little interest as there was no scientific evidence for the waters possessing healing properties. In fact, consuming the Sandrock Spring waters would have likely caused far more health issues than they supposedly cured.

Figure 1.5

Map of the chines around the island

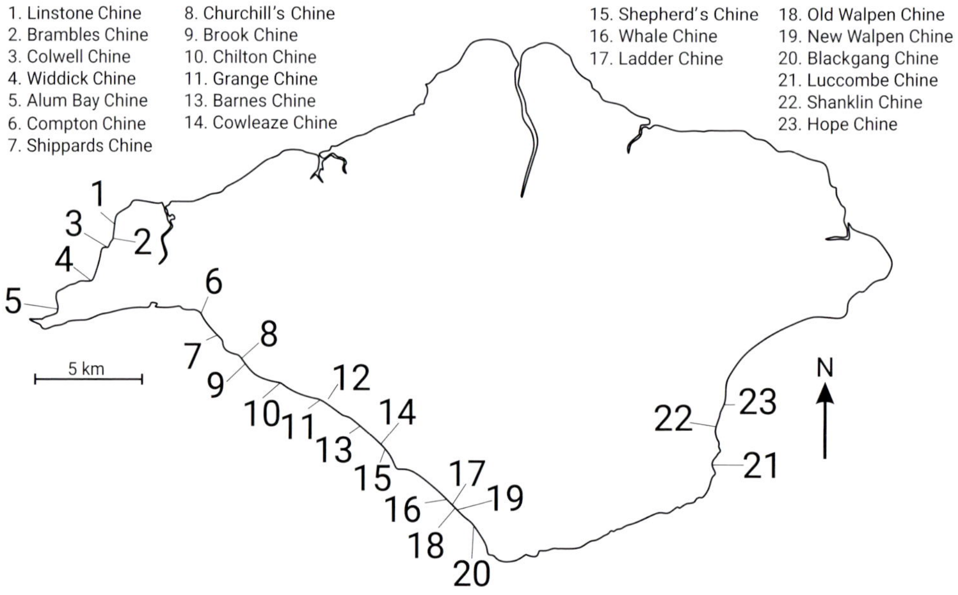

1.4 Chines

Figure 1.6

Left: 1910 photograph of Shanklin Chine

Above: photo of the chine in 2023.

Travelling around the island, you'll notice many places called 'chine', such as Shanklin Chine or Shepherds Chine. *Chine* is a local term for a steep V-shaped cutting in the cliff, made by a stream eroding rapidly through the soft sands and clays. It is thought the word is of Saxon origin, meaning deep narrow ravine. Chines are common along the southwest coast of the island, forming almost all the access points to the beaches (*Fig. 1.5*). Similar valleys in Dorset are called Bunnies.

One of the oldest chines is Shanklin Chine (*Fig. 1.6*), which started forming around 10,000 years ago. Starting as a spring in the nearby downs, the water flowed off the hills, eventually forming a substantial stream. This stream originally flowed into the Eastern Yar river in Sandown Bay. Over thousands of years, it cut an impressive 32-metre-deep gorge into the sandstone with a large waterfall at the top. This large and scenic waterfall made the chine an attraction for visitors since the 18th Century, and now forms a unique environment many animals and plants thrive in.

Figure 1.7

Location of Blackgang Chine.

Top: 1898 Ordnance Survey map showing the original location of the Blackgang Chine.

Right: Google Earth view of the current Blackgang chine landslip.

Figure 1.8

1901 photograph of Blackgang Chine.

The most famous chine on the island is Blackgang Chine, known for the oldest theme park in the UK and the incredible fossils eroding from the cliffs. The chine was located to the west of the current amusement park, forming a deep ravine through the cliff (*Fig. 1.7*). Used by local fishermen and smugglers, it was named after the black gault clay it cut into and 'gang' meaning pathway.

In 1842, Alexander Dabell, a local businessman, purchased the chine and the land above to create a romantic holiday destination. He proceeded to construct walkways and gardens through the chine itself and steps down to the beach, opening the chine to the public in 1843 (*Fig. 1.8*).

The chine has now disappeared, due to constant landslides and coastal erosion between the 1920s and 1990s. The most destructive events were in 1928, 1961 and 1994.

One of the youngest chines is Churchill's Chine, between Brook Chine and Compton Bay (*Fig. 1.9*). It only started appearing on maps in the early 1900's. This chine takes its name from a tale of Sir Winston Churchill helping Jack Seeley at Dunsbry Farm dig drainage channels in the early 1900s. Presently this is a small chine in comparison with the others along the coast, but with increased years of erosion, the chine has widened and deepened, such that it is becoming worryingly close to the road above.

1.5 Sea stacks

The highly varied geology is the underlying factor forming the island's dramatic coastline and a wide variety of unique geomorphological features carved out by coastal erosion. Among the most iconic features of the island's coast are the sea stacks forming the Needles at the western most point of the island. These have been formed by thousands of years of headland erosion from The English Channel and The Solent, as the dominant wind direction is from the southwest. Wind speeds here often reach 80 mph, and on several occasions, over 100 mph.

The Needles

The Needles is one of the most famous scenes on the island, dominating the westernmost tip of the island with the tall white

Figure 1.9

Far left: Churchill's Chine (2023)

Figure 1.10A

The Needles.

Left: An early 1700's Illustration showing the original Lot's Wife which fell in 1764

Figure 1.10B

The Needles.

Right: A photograph showing the Needles today (2023)

Figure 1.11

Freshwater Bay.

Middle Etching of Freshwater Bay taken from the Brannon (1874), showing the arch

Right Photograph of Freshwater Bay showing Mermaid Rock and Stag Rock (2023)

chalk cliffs and a series of 3 large sea stacks, with an ionic red and white lighthouse. These are the remains of a chalk ridge which once stretched across to Old Harry rocks in Dorset, some 15 miles away.

The Needles got its name from the now 4th missing stack (*Fig. 1.10A*). This tall, slender and conspicuous stack, called Lot's Wife, collapsed in a great storm in 1764. Lot's Wife Rock stood at 120 ft tall, standing way above the other 3 stacks. It's now seen as a stump in the large gap between two stacks (*Fig. 1.10B*).

The lighthouse was installed by Trinity House in 1859, by using dynamite to create a platform on the first stack for the foundations. The lighthouse was fully manned until 1994 when it was converted to run off mains electricity. There is also a network of tunnels and shafts in the Tennyson headland, from the fort at the top of the cliff descending to beach level. They were constructed in 1885 to boost the military defences of the western Solent, and includes several naturally occurring caves.

Freshwater Bay

The sea stacks of Freshwater Bay are an iconic feature of the West Wight, featuring in illustrations dating back to the 1700s, with a huge Arch Rock taking centre stage (*Fig. 1.11A*). Sadly, Arch Rock collapsed in October 1992, with Mermaid Rock being the remains of the arch. However, in November 2020, a large section of Mermaid Rock also fell into the sea during stormy weather, dramatically altering the famous rock forever (*Fig. 1.11B*). The second sea stack, closer to the bay is named Stag Rock, after a likely apocryphal tale, of a stag leaping onto the rock from the cliff to escape a hunt.

Due to the extensive and ongoing erosion of the chalk cliffs of Freshwater Bay, there are impressive cave systems on either side of the bay which can be accessed on foot during low spring tides, or by kayak/paddle boards on neap low tides.

Other sea stacks

Short-lived sea stacks also form along the coast, usually as large lumps of sandstone from the Wessex Formation and the Lower Greensand Group (*Fig. 1.12*). These are made of softer sandstones and

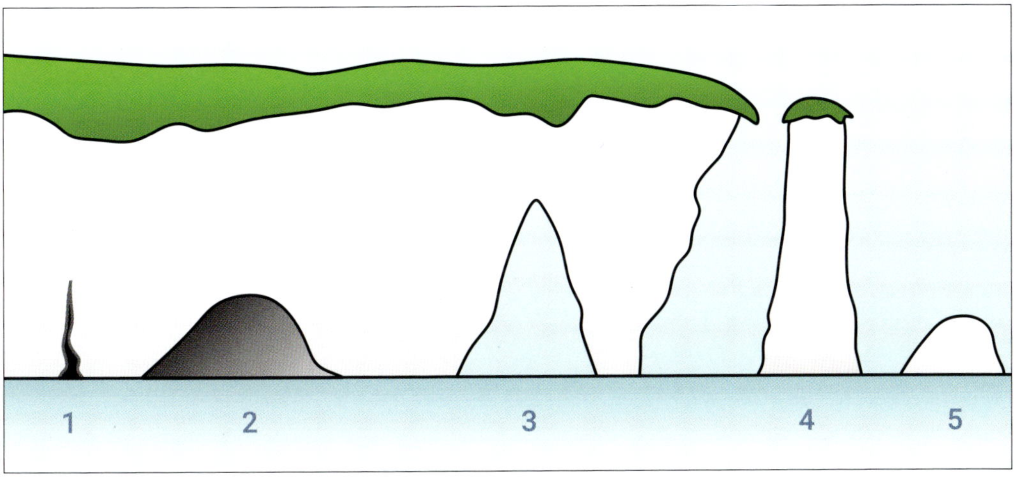

usually succumb to the sea within a year or so of formation. These stacks are not as stable and are best-given space when observing.

Figure 1.12

Left: Sea stack at Chilton Chine taken in 2023

Figure 1.13

Above: Diagram showing the 5 stages of erosion on a headland

Headland Erosion

Headlands form where a harder rock is sandwiched between softer rocks. On the island, the most prominent headlands are formed from the hard white chalk, with softer clays to the north and south. The softer clays erode faster than the chalk, creating deep bays, as at Whitecliff and Alum Bays, leaving a prominent headland of chalk jutting out into the sea.

The formation of headlands occurs in 5 stages (*Fig 1.13*).

- Stage 1. Initially starts with cracks in the rock.
- Stage 2. These cracks are expanded by processes such as waves compressing air within the crack, and salt crystallisation. This leads to incipient cave formation.
- Stage 3. The caves increase in size through wave action. Eventually, the cave erodes through the headland to form an arch.
- Stage 4. Over time, the roof of the arch is weakened by storms and weathering such as freeze-thaw, until it collapses under its own weight, forming an isolated pile of rock called a sea stack.
- Stage 5. The stack is then exposed to more weathering and erosional processes, eventually collapsing to form a stump.

After the formation of the first cave, each of these stages occur together, forming the dramatic headland seen today, such as the Needles and the Freshwater Bay sea stacks.

It seems somewhat surprising that presently there are no sea stacks at the eastern headland at Culver Down. There are, however, several caves which are called the 'nostrils' (*Fig. 1.14*). This is likely due to the fact the chalk at Culver is slightly softer than at the Needles.

Figure 1.14

Left: View of Culver Cliff taken from offshore, with the 'nostrils' visible on the bottom left-hand side of the cliff. *Below:* the 'nostrils' from the foreshore.

2. GEOLOGY

The Isle of Wight geology ranges from floodplains where dinosaurs once ruled to a deep sea teeming with life and ending with an environment similar to the Florida Everglades which was home to diverse mammals. The exposed sedimentary rocks represent over 128 million years of geological time and have been studied intensively for over 200 years (*Fig. 2.1*). Despite this prolonged and in-depth research, geologists are still finding new information every year, expanding our knowledge of the island's long geological history. The advancement of modern technology has allowed for things like paleotemperatures, age, environmental and climatic change data to be extracted from the rocks, most of which was not possible 50 years ago, due to the technology not being readily available. This chapter is a summary of the island's geology and structural history, with the current understanding of the events that formed the island. Understanding the wider geology will give you important context to the fossils you find, as well as knowing where to find specific fossils too.

Figure 2.1

The first geological map of the Isle of Wight produced by William Smith in 1815

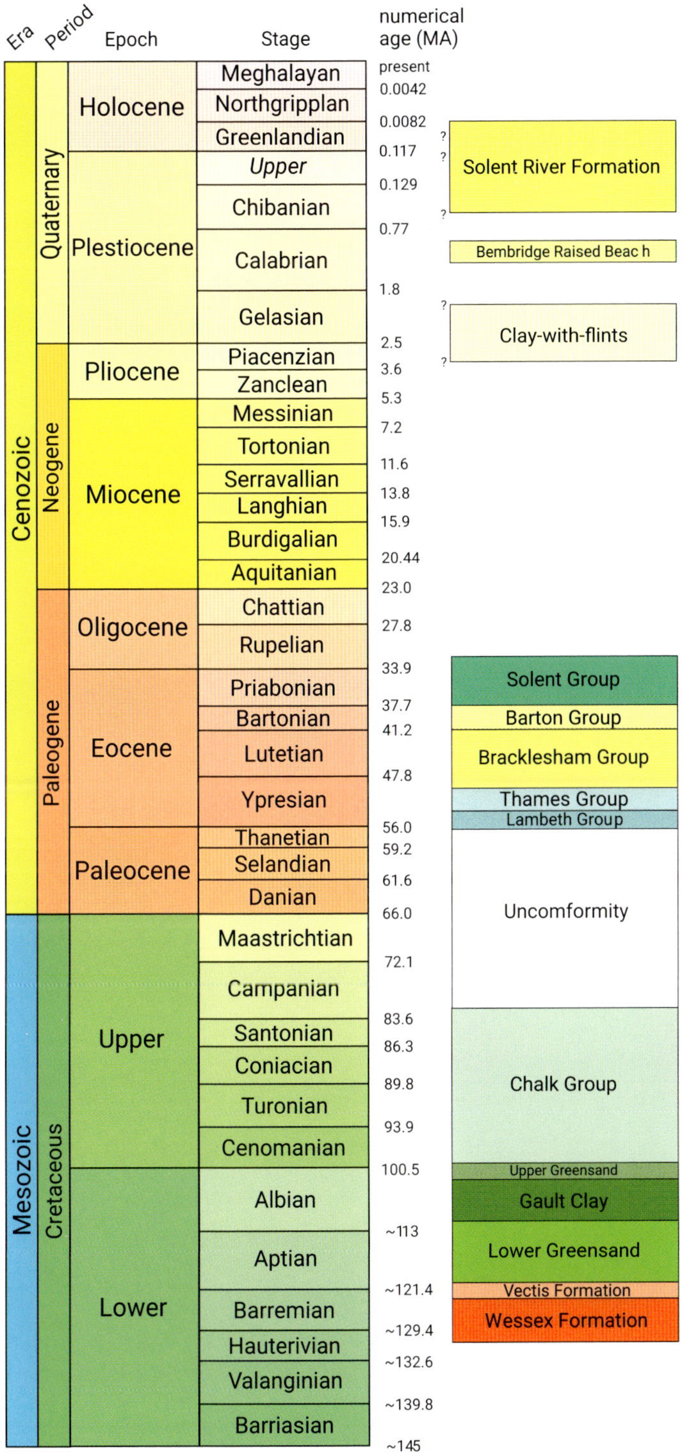

Figure 2.2

Left: Stratigraphy of Isle of Wight, with chronostratigraphic column modified from Cohen et al., 2023.

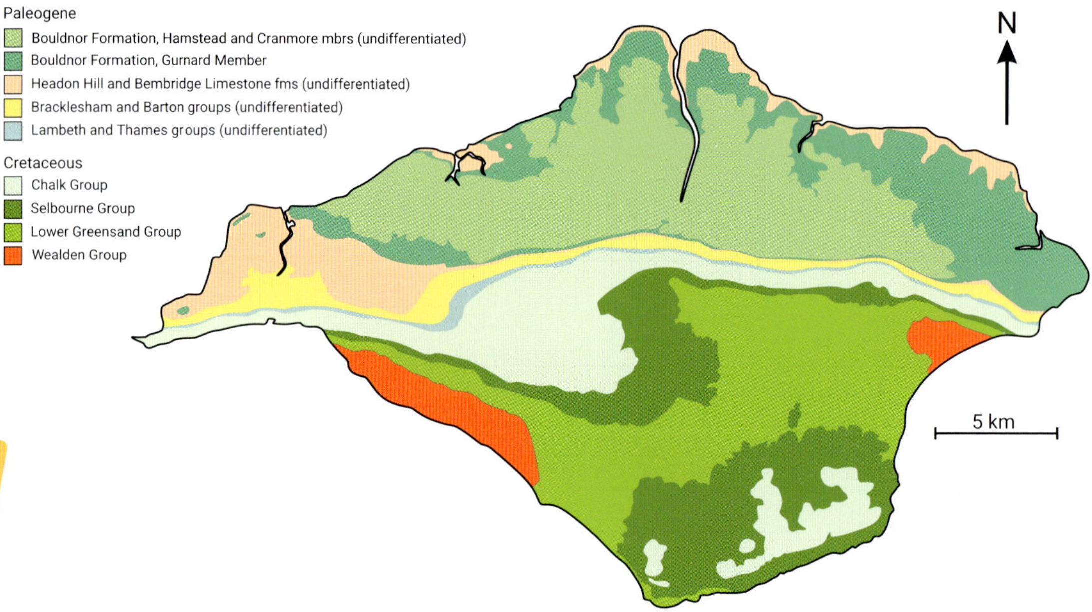

2.1 Structure

The Isle of Wight lies within a much larger geological structure, the Wessex-Channel Basin (a petroleum bearing geological area covering southern England and the English Channel).

Rocks from the Early Cretaceous to Early Oligocene are exposed beneath the remnants of a thin Pleistocene cover of ancient river gravels and soils (*Fig. 2.2*).

The oldest rocks are exposed in the south of the island and the youngest to the north. The structural history of the island is complex and similar to that of the Isle of Purbeck, Dorset, with many faults and folds creating the unique geology we see today (*Fig. 2.3*).

Figure 2.3

Top: Geological map of the Isle of Wight, modified from Gale (2019)

Figure 2.4

Right: Map showing the fault and folding structures on the Isle of Wight

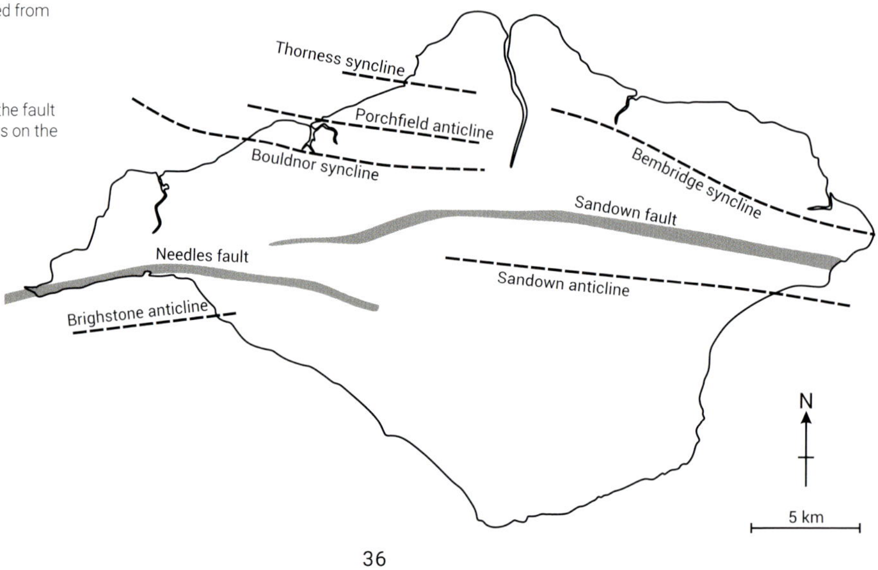

Stage 1. Normal Faulting

The structure of the island has been largely controlled by two very ancient, deep-seated faults extending east-west beneath the island, known as the Needles and Sandown faults (*Fig. 2.4*). These faults initially began moving during the Triassic (201-145 mya) through to the Early Cretaceous as normal faults, reactivated by stresses generated during the opening of the North Atlantic Ocean. This caused the strata south of the fault to be pushed down, creating the Channel Basin in which the Early Cretaceous Wealden and Lower Greensand Group accumulated (*Fig. 2.5A*). During the Early Cretaceous, Jurassic strata were exposed on the northern limb of the fault, with many fossils eroding out into the Channel Basin, now found today as reworked fossils (fossils eroded from their original geological layers into a younger sediment) in the Vectis Formation and Lower Greensand Group.

Stage 2. Hiatus

During the Albian, about 113 Mya, movement along the faults ceased. This coincided with a global sea level rise which resulted in the deposition of the Monk's Bay Sandstone, Gault Formation, Upper Greensand Formation, and Chalk formations covering the Jurassic strata in the north of the island (*Fig. 2.5B*).

After the deposition of the chalk, there was a hiatus in the sediment deposition, creating an unconformity (a time gap in the rock record). This has been attributed to the Icelandic Mantle Plume (a volcanic province – a rising column of hot magma- that raises the level of the overlying rocks) causing doming around 61 million years ago (*see page 78* for more information). At this time the top of the chalk was exposed as an erosional surface. This led to the formation of solution hollows, best seen at Alum Bay, made by prolonged dissolution of the chalk (for more about solution hollows, *see page 74*).

Stage 3. Reverse fault

During the Eocene and Oligocene, there was a north-south compression of the island's strata, which reversed the Needles and Sandown faults (*Fig. 2.5C*), leading to the formation of the Brighstone and Sandown anticlines. This event, called the Cenozoic uplift, happened in several stages, and saw the strata south of the faults uplifted, creating a basin to the north (Hampshire Basin) in

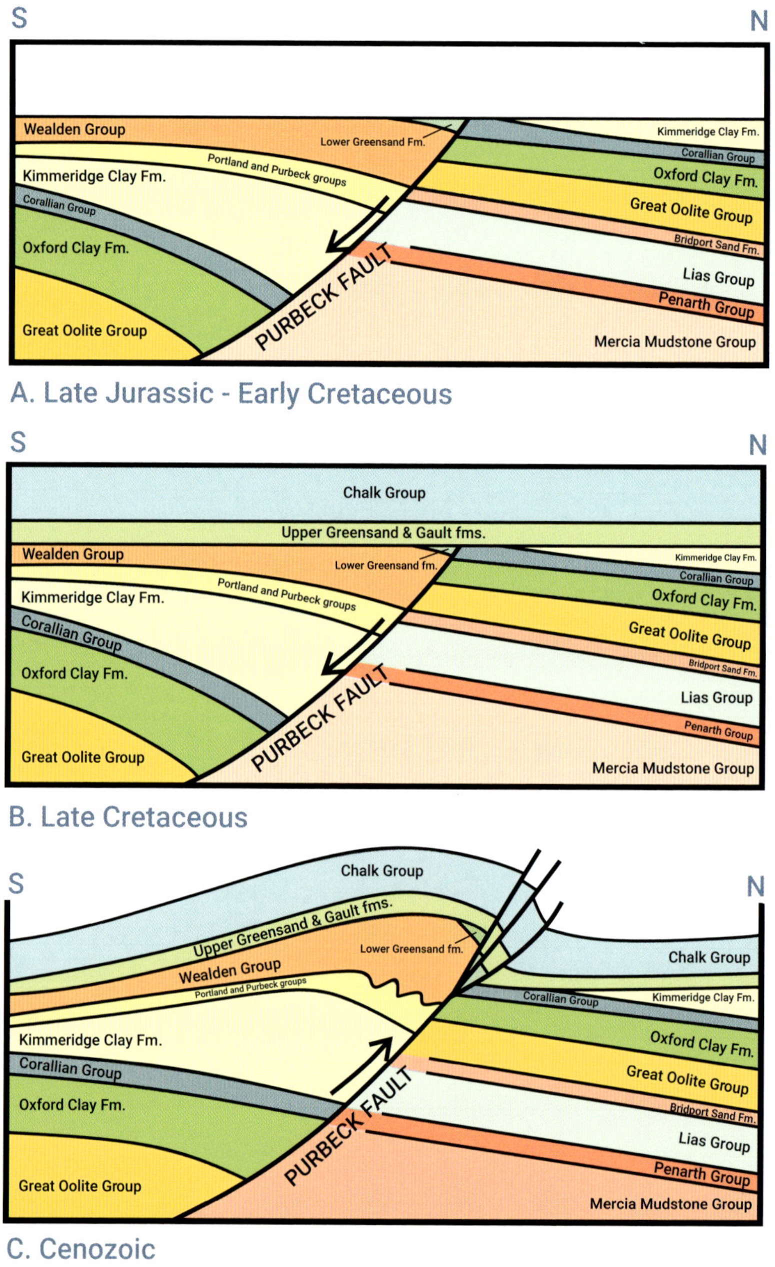

Figure 2.5

Cross-sectional diagram showing the structural history of the Needles and Sandown fault, modified from Gale (2019).

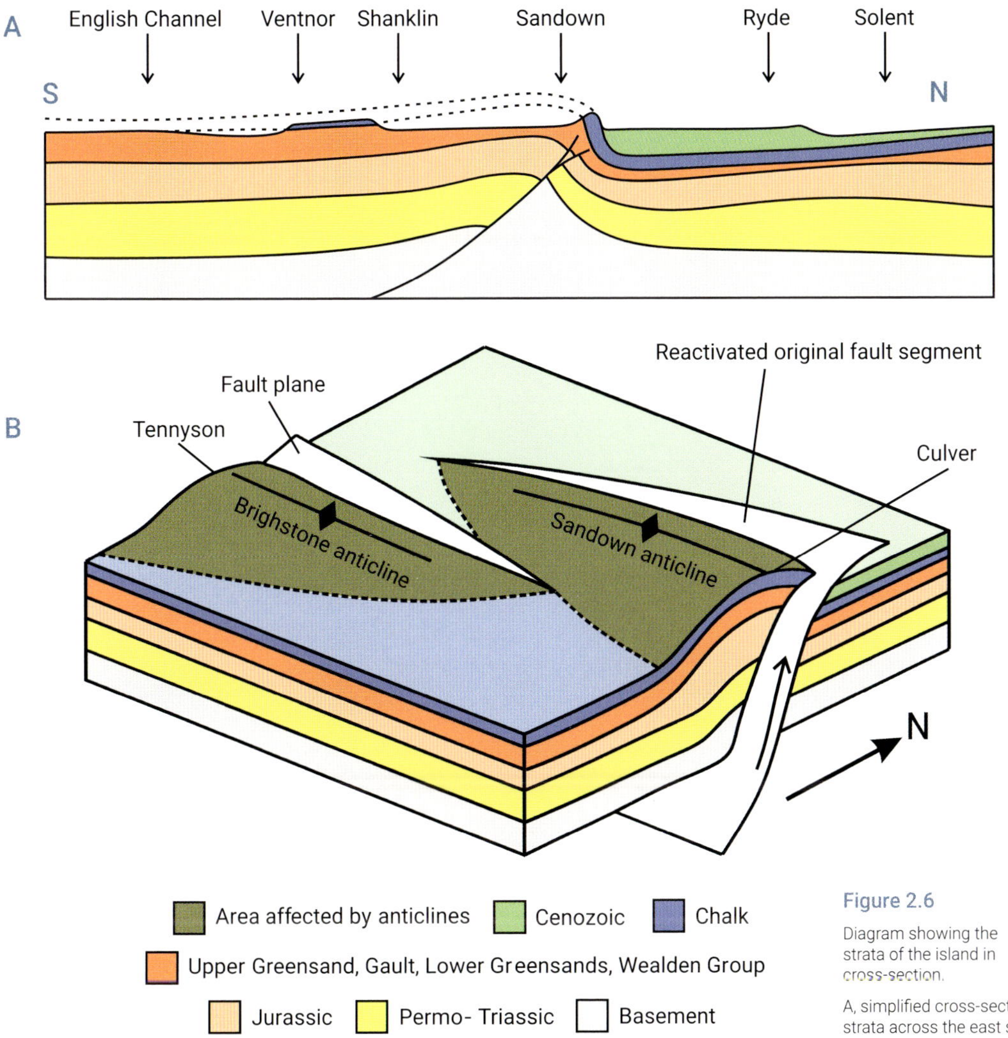

Figure 2.6

Diagram showing the strata of the island in cross-section.

A, simplified cross-section of strata across the east side of the island.

B, simplified diagram showing the Brighstone and Sandown anticlines structures across the island.

which younger Eocene strata accumulate. Due to the steep angle of the fault, sections close to the fault line are vertical or even overturned forming the Isle of Wight monocline (*Fig. 2.6*).

The Cenozoic uplift has previously been attributed to the Miocene Alpine Orogeny (uplifting forming the Alps when Africa collided with Europe) around 25 million years ago. However, evidence in reworked fossils and sediments found in Whitecliff Bay indicates the uplift occurred between 31 -34 million years ago and is associated with the Pyrenean orogeny instead.

A deep basement fault on the south side of Porchfield Harbour was also reactivated during this event, creating the Porchfield anticline and the Thorness and Bouldnor synclines (*Fig. 2.4*).

What is a fault?

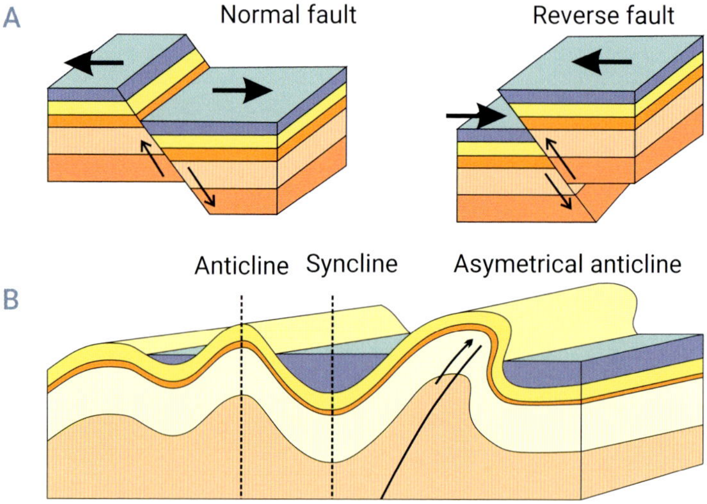

Figure 2.7

Diagram explaining A, types of faults, and B, types of folding structures found on the island.

A fault is a fracture or a zone of fractures between two blocks of rock along which movement has occurred. The fault allows the blocks to move relative to one another (*Fig. 2.7*). The movement can occur suddenly, causing an earthquake, or can occur slowly, over hundreds to thousands of years. Some faults can be active for millions of years, like the Needles and Sandown faults, and the direction of movement can change.

Fault surfaces can be at any angle. They can be on a small scale, only a few millimetres long, occurring between beds within a formation or can be on a larger scale, 10's, 100's or even 1000's of kilometres long, such as the faulting on the island or those seen between the boundaries of tectonic plates.

2.2 Milankovitch Cycles

Looking at the rocks exposed in the cliffs around the island, you may notice some very regular alternating layers, especially at the base of the chalk. These alternations of light and dark strata are the

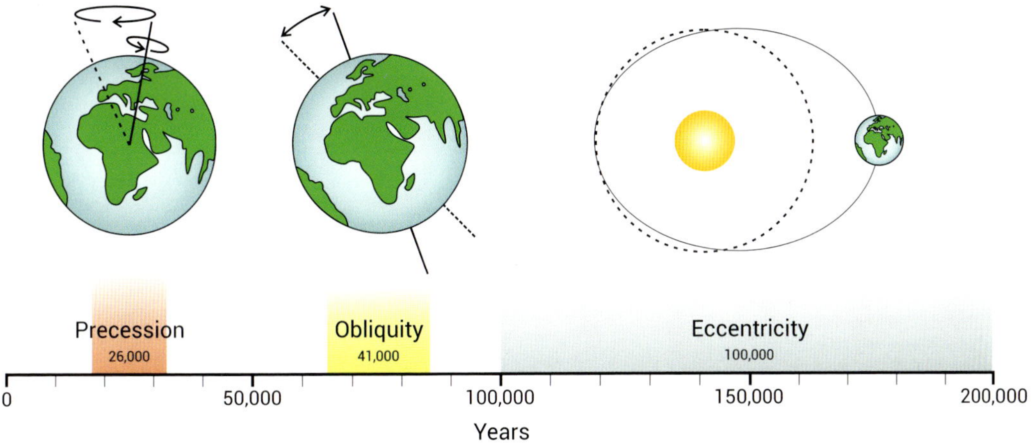

Figure 2.8

Diagram explaining Milankovitch Cycles

sedimentation reflecting larger scale cyclicity caused by fluctuations in the way Earth orbits the sun.

Milankovitch cyclicity is a combination of three dominant cycles of the Earth's rotation and path around the sun: eccentricity, axial tilt, and precession (*Fig. 2.8*). These cycles were named after Serbian astronomer and mathematician, Milutin Milankovitch, who calculated their periodicity.

- Eccentricity is the path of the Earth's orbit around the sun. The shape of the orbit changes from being a nearly perfect circle to elliptical and back again due to the gravitational fields of neighbouring planets. One complete cycle of shape change is around 100,000 years.
- Axial tilt or obliquity is the amount the Earth tilts on its axis, usually between 22.1° and 24.5°. A complete cycle of axial tilt lasts around 41,000 years.
- Precession is the Earth wobbling on its axis due to tidal forces, almost like an off-centre spinning top, and these cycles last around 26,000 years each.

The combination of these cycles changes the distribution of solar energy reaching the surface of the Earth, causing changes in seasonality, rainfall, and other climatic features. These cycles are vividly displayed in the Vectis Formation at Compton Bay and in

Figure 2.9

The Lower Chalk at Culver Cliffs showing layers of marl and chalk representing Milankovitch cycles.

the Lower Chalk at Yaverland as alternating beds of limestones and marls (*Fig. 2.9*).

2.3. Land slipping events

Landslide complex between St. Catherine's and Bonchurch

The landslide complex at St. Catherine's forms the most southerly point of the island and extends from Blackgang to Ventnor. The landslide forms an inland cliff, coining the name 'the Undercliff'. The Undercliff's landscape, with the steep St. Boniface Down sheltering the area from cold northerly winds, creates a unique warm microclimate, meaning plant species that would otherwise not survive outside the UK, thrive here. The Ventnor botanical gardens, and many Undercliff residents, take full advantage of this unique microclimate, with gardens full of warm climate species. Even banana trees are seen in many gardens.

The structure of the St. Catherine's Point portion of the landslide is best viewed (with binoculars) from Atherfield Point, where an immense block of Upper Greensand and debris form

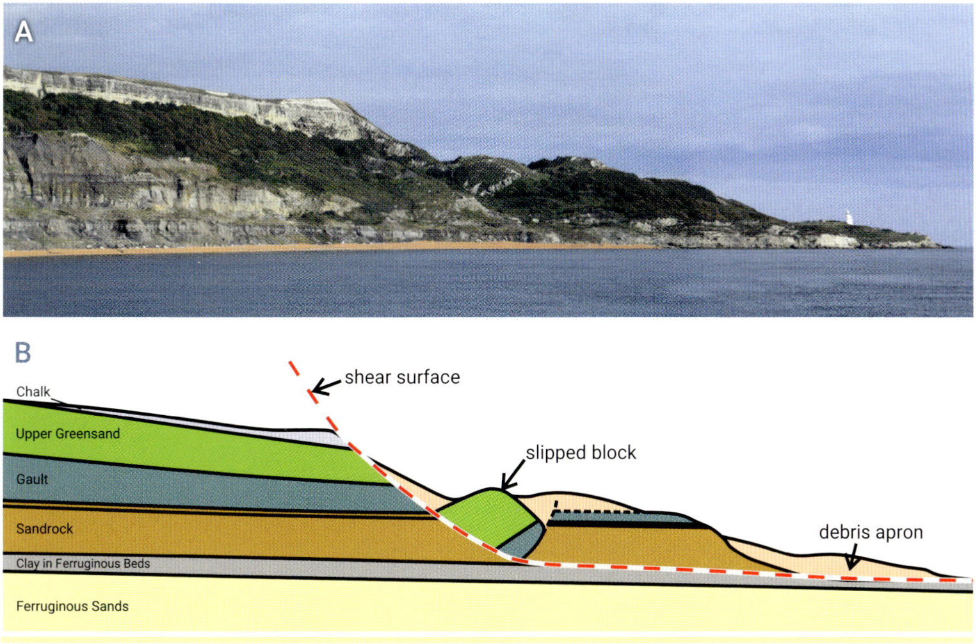

Figure 2.10

St. Catherine's landslip. A, a photo of St. Catherine's Point from the West, with B, an interpretive diagram showing the structure of the landslip complex.

several conspicuous mounds, with an apron of rock debris forming the beach (*Fig. 2.10*). The underlying Gault Clay and clay beds within the Sandrock act as slip surfaces. Large blocks of the Upper Greensand fracture and slide on the clay beneath, rotating slightly in the process as they slowly move down the slope (*Fig. 2.10*). The flanks of the undercliff are areas of highly active landslides, with major slips at Blackgang and Bonchurch ongoing (*Fig. 2.11*).

The landslide complex is thought to have been activated by the erosion of St. Catherine's Deep, a now submerged paleovalley approximately 3 km offshore of St. Catherine's Point during a time of a global sea-level rise during the Holocene. Between 9,000 and 5,000 years ago, there was a period of global warming. This caused glaciers and ice caps further north to melt, leading to a 60 metre sea level rise resulting in significant and intense erosion. This erosion created unstable conditions inland, activating the landslide.

The first landslip event occurred between Niton and Bonchurch between 8,000 – 4,500 years ago. A second landslipping phase happened between 2500 – 1300 years ago. This reactivation of the landslide was likely triggered by tree clearance during the Neolithic.

Figure 2.11

View from the Old Blackgang Road showing a large ridge of slipped strata.

However, there is evidence of Roman activity on the landslide and medieval ridge and furrow ploughing around Knowles Farm, suggesting that the landslide's significant movements predate 2,000 years ago.

Although the Undercliff area appears stable, movement data gathered to protect the lighthouse shows the debris apron moves southwards at a speed close to 10 cm a year in its lowest part, and around 2.5 cm a year in the upper portion of the landslide at St. Catherine's. Landslides continually occur from Bonchurch through to Blackgang, with a large landslide in 2021 destroying property in Ventnor. In February 2014, a large landslide occurred along the Undercliff, between Ventnor and Niton, resulting in 8 properties having to be evacuated and the road being permanently closed to traffic.

10th December 2023 saw the island's most significant landslip in over 100 years. Between Bonchurch and Luccombe, around 25 acres of land slipped towards the sea, creating a 20 metre cliff (*Fig. 2.12*). The slip destroyed property, the coastal path, woodland, and the famous Devil's Chimney, as well as causing the main Shanklin to Ventnor road to close for many months.

Between December 2023 and March 2024, significant ground movement occurred across the Ventnor – St. Lawrence area. On the 12th March 2024, a second major landslide occurred at St. Lawrence leaving 3 properties destroyed. The mass movement in the 2023/2024 winter was caused by a prolonged period of heavy rainfall.

Figure 2.12

Left: Aerial view of the 2023 Bonchurch landslip. Photo courtesy of Theo Vickers.

Below: the Smugglers Haven tea rooms are now on the edge of a 30 ft cliff. Photo courtesy of Mick Harris.

A

B

Figure 2.13

A, Compton Farm Landslip, viewed from the Military Road, with B, an interpretive diagram.

Chalk landslide at Compton Farm

Above Compton Farm is an exceptionally large, noticeable hollow on the side of the chalk ridge with an isolated hill at the bottom of the valley (*Fig. 2.13*). This is believed to be a very ancient landslip, however, there has been no formal study or dating of the large landslide to date of writing.

3. CRETACEOUS GEOLOGY

The Cretaceous geology on the island spans 50 million years, with the layers in the cliffs like pages of a book, providing an important geological record of a major sea level rise from dinosaur inhabiting floodplains to deep seas. The cliffs yield an incredible diversity of fossils indicating flourishing ecosystems evolving with the rapid environmental changes.

Geologists use a hierarchical system to categorise rocks, starting with a group, which is defined as a series of formations, representing a series of rocks deposited within a single basin, and can be traced over a large area. Formations are a distinct package of rock layers and are thick enough to be shown on geological maps. Formations can then be divided into members, which are distinct variations in the geology through the formation. Beds are the smallest recognised unit, forming distinct layers. Some beds are named, usually beds used as markers, or boundaries, but most are unnamed.

3.1 Wealden Group

Age: Lower Cretaceous, Hauterivian - Aptian (~128 – 121.4 million years)

The Wealden Group in the Wessex Basin encompasses the Wessex and the overlying Vectis formations and represents about 9 million years. Both of these formations are famous for yielding a high diversity of fossils, ranging from vertebrates, including dinosaurs, pterosaurs, crocodiles, early mammals and fishes. As well as molluscs, and insects. They also yield a diverse palaeoflora of horsetails, to ferns, cycads and conifers.

Wealden Group strata also appear in the Weald Basin of Sussex and Surrey, from where its name is derived. The Weald Basin is considered to be at least 2 million years older than the Wealden Group of the Wessex Basin on the island.

Wessex Formation

Age: Lower Cretaceous, Hauterivian - Barremian (~128 - 124 million years)

The Wessex Formation is a ~180 m thick series of sandstones and clays, spanning ~ 4 million years. These were deposited on a valley floodplain, in and around a large meandering river flowing from the west across southern England and into northern France (*Fig. 3.1*). The source of the river is unknown but often said to be in the Cornwall/Devon area, but offshore deposits indicate it may have been in the southern Irish Sea. This river likely drained into the Neo-Tethys Sea, just south of where Paris is today. The formation is internationally famous for its high diversity in fossils, including dinosaurs, micro vertebrate fossils (mammals, lizards, amphibians), and even insects in amber.

Paleosols

Most of the Wessex Formation consists of red and mottled clays, called paleosols, which are fossilised soils. They occur in a variety of colours, from red, purple, blue, green, orange, and yellow (*Fig. 3.2*).

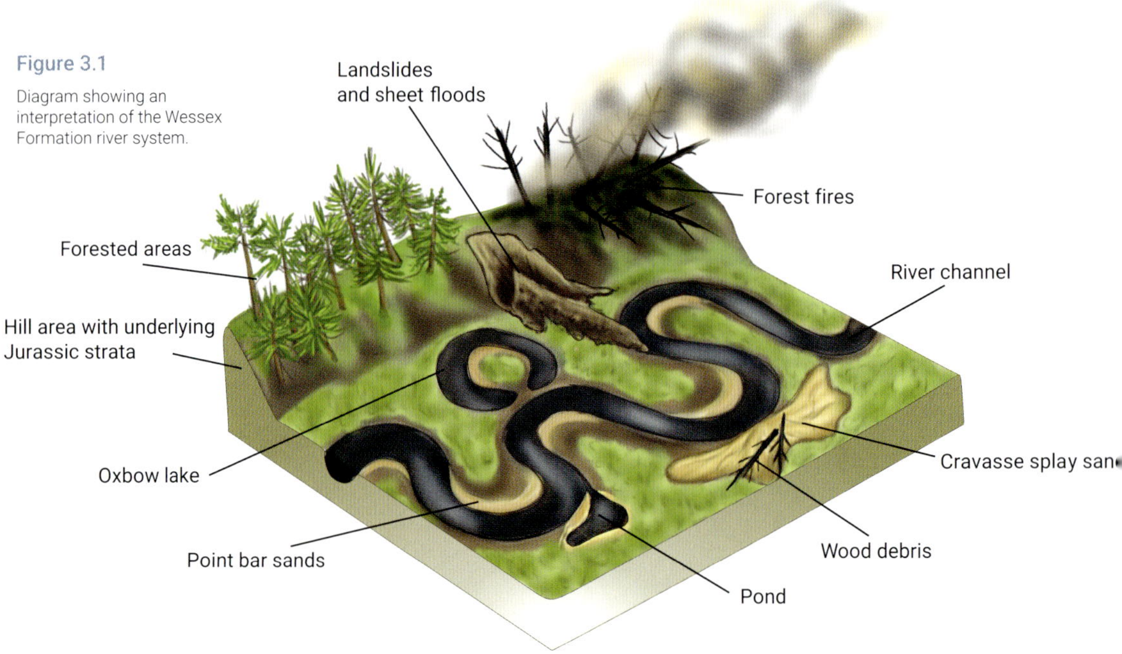

Figure 3.1

Diagram showing an interpretation of the Wessex Formation river system.

Figure 3.2

Exposure at Grange Chine shows colourful mottled paleosols and an overlying plant debris bed.

The colours of the paleosols are dictated by soil forming processes, with wetting and drying and fluctuating water tables, along with the introduction of reactive organic matter from forests growing on the floodplain. These paleosols are a crucial key in understanding ancient environments, and preserve a vast amount of information, as the type of soil is dictated by a combination of factors, such as the climate, proximity to a water source, ground water chemistry, and the type of plants growing in the area.

The red clays are formed under dry conditions, indicating iron in the soil was exposed to the air for a prolonged period, allowing it to fully oxidise. The mottled soils are caused by fluctuations in ground water levels and chemistry. Blue-grey clays with complex but irregular patterns are a result of both the groundwater and seasonal rains entering the soil, causing the iron compounds to reduce. Purple and green mottled clays are depleted in iron (Fe) but have more ferrous iron (Fe^{2+}).

Sandstones

There are numerous sandstone units through the Wessex Formation, varying in lateral extent and thickness. The largest and most notable is the Sudmoor Point Sandstone, which crops out over a mile of cliff and is around 6 metres thick, and is described as a point bar

Figure 3.3

Exposure of the Sudmoor Point Sandstone, looking west from Chilton Chine.

sandstone, deposited on the inside bends of the Wessex Formation river (*Fig. 3.3*). There is a distinctive conglomerate at the base, locally known as the 'popcorn bed', which is known for yielding well-preserved dinosaur teeth and the occasional bone (*Fig. 3.4*).

There are also many thin lenses of sandstone in the Wessex Formation, which are interpreted as small banks of sand from river inlets or during flooding events (*Fig. 3.5*).

Another notable sandstone is the Hanover Point Sandstone, which is likely a crevasse splay sandstone, formed when the river burst its banks. The bottom of the sandstone famously yields very large sandstone-infilled dinosaur foot casts, and the top portion

Figure 3.4

The 'popcorn' bed conglomerate at the base of the Sudmoor Point Sandstone, showing angular rip-up clasts of clay and small lumps of caliche (a natural cement of calcium carbonate and other materials such as clays or sands).

Figure 3.5

The clay-infilled erosional channel within the Brighstone Sandstone. Willing student for scale.

is heavily burrowed. This top portion represents the remains of a standing body of water, that was eventually infilled with plant debris by a flooding event (*Fig. 3.6*).

Plant Debris beds

Plant debris beds, or PDB's are found throughout the formation but they are not laterally continuous. They are, however, highly fossiliferous, noted for a high abundance of plant material, and frequently yielding skeletons of dinosaurs and other vertebrates (*Fig. 3.7*).

There are two hypotheses regarding plant debris bed formation which are widely accepted:

> **Hypothesis 1** by Sweetman and Insole, 2010. The Wessex floodplain was an incredibly challenging environment for wildlife, with periods of extreme drought and wildfires, followed by heavy rainfall causing flash flooding and large debris flows. The debris flows picked up anything and everything on the floodplain in its path, including the burnt

Figure 3.6

Top: The Hanover Point Sandstone at Hanover Point.

Right: a bioturbated dorsal surface.

logs from forest fires, as well as remains of animals. This includes teeth, bones and partial to complete carcasses. Some of these debris flows were deposited in ponds and depressions on the floodplain.

Hypothesis 2 by Wright et al., 2000. The Wessex floodplain was a complex environment, with frequent wet and dry periods and periodic wildfires. The low-relief areas were mainly submerged or partially stagnant ponds or billabongs. The water levels in the ponds were dictated by seasonality and the rise and fall of the water table.

Figure 3.7

Top: An exposure of a plant debris bed in a slumped portion of cliff.

Left: Large logs eroding from a plant debris bed at the top of the cliff.

During the wet seasons, these low relief areas acted as a catchment for sediment during flooding events, which deposited the remnants of wildfires, along with animal remains. Crocodiles, turtles and fishes occupied these 'ponds' using them as refuges during the dry season, like 'gator holes' seen in the Florida Everglades today. This explains the concentration of vertebrate remains and plant debris.

Both hypotheses have good evidence to support them, however, the formation of plant debris beds is likely to be complex, involving a combination of flooding events and 'gator' holes. Either way, they yield a high number of vertebrate fossils, commonly isolated teeth and bones, and on rare occasions, partial to near complete skeletons.

Vectis Formation

Age: Lower Cretaceous, Barremian - Aptian (124 – 121.4 million years)

The Vectis Formation is a series of clays, laminated mudstones, and nodular limestones spanning less than 3 million years. These strata were deposited in and on the margins of mainly shallow freshwater and brackish lagoons. However, there were short-lived marine horizons throughout the formation, which are rich in oysters and other marine bivalves. These marine incursions heralded the start of a major global sea level rise. The Vectis Formation on the Isle of Wight consists of 3 members: the Cowleaze Chine, Barnes High Sandstone, and the Shepherd's Chine members.

The Cowleaze Chine Member consists of 7 - 10 metres of finely laminated dark grey mudstones and light grey siltstones and sandstones. The base of the member is marked by a 1-metre thick white sandstone, called the 'white rock', which is used as the boundary marker between the Wessex and Vectis formations (*Fig. 3.8*). The base of the member is full of the brackish water bivalve *Filosina* changing into beds abundant with the freshwater gastropod *Viviparus*. It is thought to have been deposited in a shallow lagoonal or mudflat environment.

The Barnes High Sandstone Member is one or more thick yellow/brown/buff sandstones with laminated mudstones in between, reaching a thickness of around 2.5- 7 metres thick depending on the location. The thickest and most extensive exposures of the sandstone are at Barnes High on the southwest coast, but it is seen at Compton Bay and Yaverland (*Fig. 3.9*). The sandstone contains many trace fossils and often displays cross-bedding. The very top of the sandstone consists of a thin mud clast conglomerate with freshwater bivalves such as *Unio* and the brackish water bivalve *Filosina*, mostly pyritised. There have been many suggestions for the depositional environment of the Barnes High Sandstone Member, but it is now widely thought that it was a prograding (building up towards the sea) lagoonal delta, subject to wave action on its shoreline.

Figure 3.8

Right: A, the Wessex-Vectis boundary at Compton Bay, with the white rock boundary indicated with a dotted line.
B, the white rock at the base of the cliff with a lower dinoturbated surface.
C, bioturbation in the sandstone.
D and E, three-toed dinosaur foot casts loose on the beach.

A
B
C
D
E

Figure 3.9

Top: the Cowleaze Chine and Barns High members at Compton Bay.

Bottom: the Barns High Sandstone forms a prominent bed at the top of the cliff, at Barns High, Brighstone.

The Shepherd's Chine Member is a ~50 metre thick series of rhythmically bedded fine-grained sands, silts, and mudstones with alternating light grey fine sand/silt to a dark grey mudstone. In the highest parts of the member, there are several thin limestones, packed with the bivalve *Filosina*, while near the top, the oyster *Liostrea* becomes common where it forms dense shell beds (*Fig. 3.10*). Some of the limestone beds are storm generated coquinas, with the waves stirred up the lagoon floor and concentrating the shells. These beds consist of broken and jumbled *Filosina* bivalves and often contain bony fish, sharks, and rarely, pterosaur remains. Rare dinosaur footprints in these shelly limestones show that they formed

Figure 3.10

Filosina limestone slabs.

Top: At Compton Bay covering the top of the beach.

Bottom: Slabs showing desiccation crack infills.

in very shallow water. Some of the dinosaur footprints are infilled with shells of the freshwater gastropod *Viviparus* sp. and may show desiccation cracks on the underside. These indicate fluctuating water levels throughout the member and show that this member was deposited in a very shallow, storm-influenced lagoon.

The fossils indicate fluctuations in salinity through the member; with freshwater unionid bivalves in some horizons and brackish *Filosina*, oysters, and ostracods in others. Between Shephard's Chine and Atherfield Point, the member contains numerous pyrite nodules. These nodules accumulate on the foreshore and may contain fish scales, and rare, pterosaur bones. Marine reptile remains have also been found in the brackish horizons, including the plesiosaur *Vectocleidus pastorum*, showing that the lagoon must have been open to the wider sea in some places.

3.2. Lower Greensand Group

Age: Lower Cretaceous, Aptian – Albian (121 – 112 million years)

The Lower Greensand Group on the island is famous for yielding a high diversity of marine shelly fossils including corals, ammonites, bivalves and brachiopods. Perhaps the most impressive fossils are the giant heteromorph ammonite, *Tropaeum* from Chale Bay (*see page 191*). The group comprises a series of clays and sands deposited in estuarine to open marine environments and includes four formations: the Atherfield Clay Formation (*Fig. 3.11*), Ferruginous Sands Formation, Sandrock Formation, and the Monk's Bay Sandstone Formation (formerly known as the Carstone).

The name 'Greensand' is coined from the green colour of the beds when fresh. This greenish tint is due to the presence of the mineral glauconite, a green iron silicate mineral, forming only in shallow marine conditions. It is formed by authigenesis, a process whereby the mineral forms by crystallisation within the sediments, rather than having been transported from elsewhere.

The upper part of the Ferruginous Sands Formation at Yaverland contains thin conglomerates with derived Upper Jurassic fossils. These likely come from the Oxford and Kimmeridge Clay formations and include fragments of ammonites, occasional crocodile teeth and rarer marine reptile bones.

Atherfield Clay Formation

The Atherfield Clay Formation is a series of grey to brown mudstones and was deposited in open marine conditions. It is between 20-50 metres thick and consists of 5 distinct units: Perna Bed Member, Chale Clay Member, Lower Lobster Bed Member, Crackers Member and Upper Lobster Bed Member.

Figure 3.11

Outcrop of Lower Greensand Group at Atherfield Point.

The Perna Bed Member marks the boundary between the Wealden Group and the Lower Greensand. It is a very prominent 1.5m thick, hard sandstone, and being harder than the surrounding rock, it protrudes slightly in the cliff forming an obvious ledge (*Fig. 3.12*). The Perna Bed Member was deposited in a shallow marine shelf environment, representing the early Aptian marine transgression.

At Atherfield Point, the Perna Bed Member forms a prominent ledge called The Bench, which extends over half a mile out to sea at Atherfield Point. The Bench has been the cause of hundreds of shipwrecks before modern navigational equipment came into use. Highly fossiliferous blocks of the Perna Bed are often found littering the beach between Shepherds Chine and Atherfield Point, and at Yaverland. Most notable among the fossils are the thick-shelled large oysters *Aetostreon latissima* and the rounded masses of the compound coral *Holocystis*. The base of the Perna Bed Member is a fine conglomerate, containing reworked Jurassic fossils, including fragments of ammonite, pieces of silicified oysters and rare vertebrate remains.

The Chale Clay Member is an 18 – 19 metre thick unit of brown to grey clay, outcropping at Atherfield Point. Fossils are rare here,

A
B
C
D

but it has yielded ammonites, echinoids, bivalves, crustaceans, and fish remains. More commonly found are red-brown sideritic nodules, and some are *Favreina* pellet-lined crustacean burrows called *Ophiomorpha*.

The Lower Lobster Bed Member is an 11.6 metre thick blue-grey unit of mudstones and silty clays. Due to the south-eastward dip of the strata, the lobster bed outcrops on the foreshore, some 550 - 600 metres east of Atherfield Point, allowing for easy fossil collecting (*Fig. 3.13*). The member gets its name for famously yielding a high abundance of exceptionally well-preserved, articulated examples of the lobster-like decapod shrimp, *Atherfieldastacus* (formally called

Figure 3.12

Far left: Perna Bed.

A, fallen loose blocks on the beach at Atherfield. B, one of the blocks full of shell fossils. C, the Perna Bed on the foreshore at Atherfield Point, forming a prominent ledge extending out to sea. D, the base of the Perna Bed at Yaverland composed of Jurassic oyster fragments and other reworked fossils and detrital grains. Scale equals 10mm.

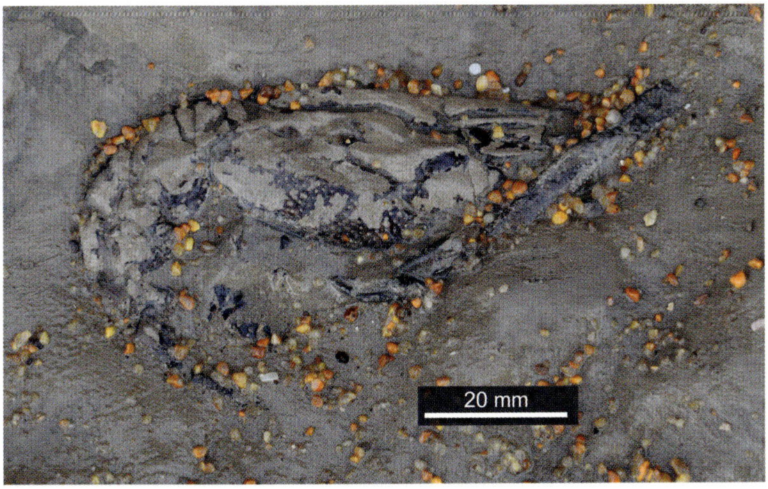

Figure 3.13

Top: Lower Lobster Bed and Crackers Member exposure at Atherfield

Left: An in situ lobster.

A
B
C

Meyeria). These so-called 'lobsters' are often found in the 'classic' shrimp pose with the telson (tail) tucked underneath the body. A few specimens retain the original blue colour of the carapace. A calcite-infilled ammonite, *Deshayesites*, is also a common find here. The Crackers Member is a dark blue–brown, around 6-metre thick sandstone which is very prominent in the west of Chale Bay (*Fig. 3.14*). The member contains large concretions with fossiliferous cores, for which the member gets its name. There are two types of 'crackers', hard and soft. Hard crackers usually are very hard concretions with unfossiliferous cores. Soft crackers have a sandy core, which can preserve a high diversity of fossils, including bivalves, brachiopods, brittle stars and ammonites (*Fig. 3.14*). The soft crackers are highly sought by fossil collectors.

Figure 3.14

Far left: A, exposure of the Crackers Member showing large cracker nodules protruding on the surface. B, a soft cracker with well-preserved shells in the core. C, large fallen cracker with a core of the bivalve *Gervillia* sp. Scale equals 5mm.

The Upper Lobster Bed Member consists of 14.5 metres of alternating dark grey mudstone and a lighter grey sandstone. Flat nodules can be found in the lower beds, however, fossils here are scarce compared to the Lower Lobster Bed.

Ferruginous Sands Formation

The Ferruginous Sands Formation is around 161 metres thick, and outcrops at Atherfield through to Blackgang, as well as Compton Bay and at Yaverland where it forms Red Cliff (*Fig. 3.15*). This formation makes up much of the inland outcrops of the Greensand Group and can also be seen in many road cuttings across the southern half of the island.

The formation has been divided into 10 units, with the beds numbered (IV-XIV), as well as given member names: IV (Lower Gryphaea Beds), V (Scaphites Beds), VI (Lower Crioceras Bed), VII (Whale Chine Member), VIII (Upper Crioceras Bed), IX (Ladder Chine Sands), X (Upper Gryphaea Bed), XI (New Chine), XII (Foliated Clay and Sand), XIII (Walpen Chine Member) and XIV (Blackgang Chine Member). The two terms for each member are used interchangeably amongst local collectors.

The Gryphaea beds are famed for yielding a high abundance of the large oyster *Aetostreon* (in the family Gryphaeidae, hence the name), bryozoan colonies and clusters of brachiopods called 'nests'

Figure 3.15

Top: Exposure of the Ferruginous Sands Formation at Yaverland viewed from offshore.

Right: Whale Chine showing VIII (Upper Crioceras Bed), VII (Whale Chine Member) and VI (Lower Crinoceras Bed) members and fallen blocks of VIII and VI members within the chine.

(*Fig. 3.16*). Beds V and VI yield the ammonites *Australiceras*, *Tropaeum* and *Cheloniceras* which can reach considerable sizes, where only the most ardent collect such heavy monsters. *See Figure 3.17* for the most recent specimens collected by Jack Wonfor (no one knows how he managed to lift them). Beds VII – IX commonly yield phosphatised cores containing ammonites and a diverse fauna of molluscs and brachiopods.

Figure 3.16

Far Right: A, Lower Gryphaea Bed forming a ledge in Chale Bay with B, a large *Aetostreon* oyster with a bryozoan colony.
C, large *Aetostreon* oyster clusters and lots of trace fossils in fallen blocks.

Sandrock Formation

The Sandrock Formation is a ~70 metres thick bed of yellow sands with interbedded clay bands, and thick clay beds. It outcrops at Blackgang Chine and Shanklin, as well as several inland sites. It is

A
Lower Gryphaea Bed

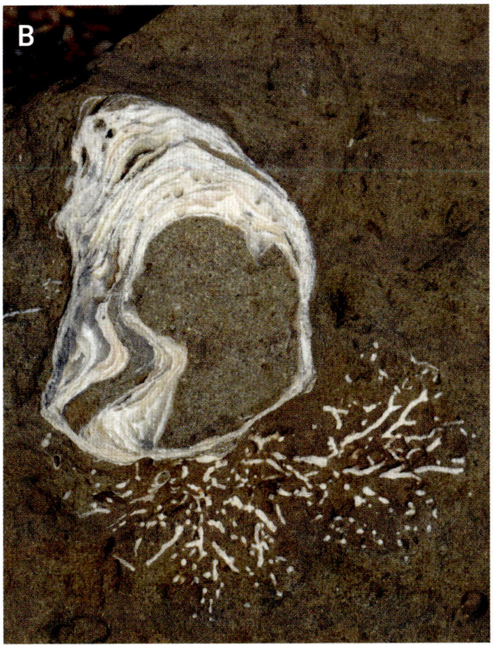
B

C

Figure 3.17

Jack Wonfor with his giant ammonites.

A, *Ammonitoceras* (October 2023) and B, *Caspianites* (May 2023).

thought to represent an estuarine environment with subtidal channels. The formation yields bored fossil wood in abundance and exceedingly rare ammonites.

Monk's Bay Sandstone Formation

The Monk's Bay Sandstone Formation, formerly known as the Carstone, ranges from 2 – 20 metres thick. The formation consists of a brown, coarse sandstone, containing quartz gravels and other pebbles. It also yields phosphatic concretions containing ammonites amongst other marine fossils.

3.3. Selbourne Group

The Selbourne Group is a suite of strata with a thickness of up to 110 metres. It is divided into two formations, a lower Gault Clay Formation overlain by the Upper Greensand Formation. The boundary of the two formations is marked by the so-called 'Passage Beds', a dark silty to sandy marl that contains carbonate concretions.

Gault Clay Formation

Age: Lower Cretaceous, Albian (109 – 103 million years)

A 30 – 40 metre thick, dark grey silty clay, with thin sandy glauconitic horizons. On the mainland (Kent and Sussex), this formation is highly fossiliferous, but on the island Gault Formation fossils are rare, with pyritic and phosphatic ammonites being the most commonly found Gault fossils. The Gault crops out at Compton Bay, Blackgang, St. Catherine's, Luccombe and Yaverland (*Fig. 3.18*). It was deposited in an offshore marine environment, with water depths of 60 metres where the bottom waters are largely undisturbed by wave action. On the island, the Gault Clay is often the cause of landslides, hence it is locally known as 'blue slipper'.

Upper Greensand Formation

Age: Lower Cretaceous, Albian (102 – 100 million years)

The Upper Greensand Formation is up to 75 metres thick and consists of green to grey calcareous sandstones, with varying amounts of glauconite and mica grains. There are carbonate concretions and black phosphate pebbles scattered through the formation but concentrated in its upper two-thirds. The upper portion of the formation contains nodules of black chert and silicified concretions. The chert beds cap the impressive Gore Cliff above the Blackgang Chine amusement park (*Fig. 3.19*), and the succession can be seen up close along Cripples Path

Figure 3.18A

Gault exposure between Bonchurch and Luccombe.

Figure 3.18B

Gault exposure at Yaverland.

at Niton Undercliff (*Fig. 3.20*). It is regarded to be an offshore shelf environment (shallow, flat lying seabed extending from the continent's coastline, less than 200 metres deep) with lower shoreface horizons due to a prograding shoreline (where the shores begin to build seawards).

3.4. Chalk Group

Age: Upper Cretaceous, Cenomanian – Campanian (100 – 78 million years)

The white chalk forms the most dramatic coastal scenery on the island. With the famous Needles in the west and the steeply dipping cliffs of Culver Down in the east, and a chalk ridge that forms the 'backbone' of the Isle of Wight. The Chalk Group can be traced across southern England, from the famous white cliffs

Figure 3.19A

Gore Cliff exposure showing the bedded Upper Greensand Formation. Taken from the Old Blackgang Road carpark at Windy Corner. The site of a large rockfall in 1928, which destroyed the coast road forms a debris pile from this rockfall, marked by the white arrow.

Figure 3.19B

The bedded chert beds of the Upper Greensand Formation.

of Dover to Old Harry's Rocks at Swanage and to the almost vertical cliffs forming Lulworth Cove in Dorset. It also forms the North and South Downs, Salisbury Plain, and Chiltern Hills and extends through Cambridgeshire and the wolds of Lincolnshire and Yorkshire. On low spring tides, the stretch of coastline from Yaverland to Whitecliff Bay boasts an almost complete succession of

Figure 3.20

Cripples path at Niton Undercliff, where it cuts underneath the Upper Greensand overhangs.

the group (no gaps in the geology), missing only the last 20 million years of the Cretaceous.

The Chalk Group represents a period of a major sea level rise when levels were up to 200 metres higher than today. Much of the sea level rise was tectonically induced by the development of mid-ocean ridges in the Atlantic Ocean. The melting of polar ice due to a rise in global temperatures may also have contributed to the sea level rise. During the deposition of the chalk, low-lying areas were submerged and turned into shallow seas. As a consequence, the continental shelves became deeper marine environments and were starved of sediment derived from the land, leading to the production of almost pure biogenic limestones.

The Chalk Group is one of the most studied geological stratigraphical units in England, due largely to the laterally extensive coastal outcrops that continue inland within quarries, roadways, and natural erosional bluffs, numbering over 300 localities in southern England. There are many quarries on the island, excavated for lime by farmers to help neutralise acid soils.

Grey Chalk Subgroup

The Grey Chalk ranges from 40 – 60 metres thick and is composed of marls and marly limestones, with rhythmic bedding likely the result of Milankovitch cyclicity (*Fig. 3.21*). The lower portion of the Grey Chalk Subgroup is clay-rich, while the upper portion is more calcareous, leading to its division into two formations - West Melbury Marly Chalk and the Zig Zag Chalk formations.

The base of the West Melbury Marly Chalk Formation is known as the Glauconitic Marl Member, which is highly fossiliferous, yielding ammonites, bivalves, brachiopods, echinoids, sponges and the occasional vertebrate remains (*Fig. 3.22*). Boulders of the Glauconitic Marl Member can be found on the foreshore at St. Catherine's Point, readily identified by the greenish colour and dark brown phosphatic nodules on the surface, making for easy fossil collecting. The Zig Zag Chalk Formation also yields ammonites, bivalves, brachiopods, and echinoids.

White Chalk Subgroup

The White Chalk Subgroup is complete within the island's stratigraphy and comprises of 7 distinct formations: Holywell Nodular Chalk, New Pit Chalk, Lewes Nodular Chalk, Seaford Chalk, Newhaven Chalk, Culver Chalk and the Portsdown Chalk formations in ascending order.

Figure 3.21

Grey Chalk Subgroup exposure at Yaverland.

Figure 3.22A

Exposure of the West Melbury Marly Chalk Formation at Yaverland.

Figure 3.22B/C

Above: Prominent carbonate concretions.

Right: Most of these have originally silicious sponges preserved within.

Holywell Nodular Chalk Formation

Age: Upper Cretaceous, Cenomanian – Campanian (100 – 78 million years)

The Holywell Nodular Chalk Formation is 30 metres thick of nodular chalk with thin marl beds. The lowest most bed is called the Plenus Marls named after the belemnite *Actinocamax plenus*. It is around 3 metres thick, and comprises marls and marly chalks. It is rather grey which contrasts with the surrounding white chalk beds. The formation has an abundance of inoceramid bivalve fragments, however, the belemnites after which its named are very rare.

New Pit Chalk Formation

Age: Upper Cretaceous, Cenomanian – Campanian (100 – 78 million years)

The New Pit Chalk Formation is a 25 metre unit of pure white hard chalk, with a low abundance of fossils, although brachiopods can be common. The formation shows a weak rhythmic sequence of alternating marls and chalk, with rare flints. A road cutting along the military road between Afton to Compton road is the most easily accessible outcrop of this formation.

Lewes Nodular Chalk Formation

Age: Upper Cretaceous, Cenomanian – Campanian (100 – 78 million years)

The Lewes Nodular Chalk Formation is 40-45 metres of interbedded chalk with alternations of hard nodular chalk and soft chalky marls. It is rich in black flint nodules, many of which have formed within invertebrate burrows. It is richly fossiliferous with echinoids, bivalves and brachiopods. This formation is exposed and accessible along the road cutting on the Military Road, in Freshwater Bay and less easily accessible at Watcombe Bay and Culver Cliffs. There are other exposures in road cuttings and quarries across the island. This formation forms the hill on which Carisbrooke Castle is situated.

Seaford Chalk Formation

Age: Upper Cretaceous, Cenomanian – Campanian (100 – 78 million years)

The Seaford Chalk Formation is 73 - 95 metres of soft white chalk with evenly spaced seams of nodular flints. This formation is best seen in Freshwater Bay and Scratchell's Bay (*Fig. 3.23*).

Newhaven Chalk Formation

Age: Upper Cretaceous, Cenomanian – Campanian (100 – 78 million years)

The Newhaven Chalk Formation is around 100 metres of soft to hard chalk, with marl seams and bands of flint. The full succession of the formation is seen in Scratchell's Bay and Culver Cliff.

Culver Chalk Formation

Age: Upper Cretaceous, Cenomanian – Campanian (100 – 78 million years)

The Culver Chalk Formation is around 100 metres of soft white chalk, with well-developed nodular flint beds. With exposures at

Figure 3.23

Seaford Chalk Formation.

A, exposure at Freshwater Bay, with B, in situ flint nodule, C, periglacial (freeze thawed) shattered flints, and chalk.

Scratchell's Bay and various chalk pits across the island, the best exposure is in Whitecliff Bay where it yields echinoids.

Portsdown Chalk Formation

Age: Upper Cretaceous, Cenomanian – Campanian (100 – 78 million years)

The Portsdown Chalk Formation is 63 metres thick and is the youngest chalk on the island. There are flints throughout and marly layers. The section is seen at Scratchell's Bay, Culver Down and Shide Quarry amongst others.

3.5 Solution hollows

The top of the chalk, best seen at Alum Bay, has large cavities over its surface (*Fig. 3.24*). These are called solution hollows, formed by

Figure 3.24

A, solution hollows in the top of the chalk at Alum Bay.

B, showing the flint debris infill.

prolonged chemical weathering of the rock. It was caused by slightly acidic rain and groundwater dissolving away the carbonate-based chalk below, creating sinkholes.

Solution hollows in other areas can act as a catchment for fossil vertebrate remains, as animals fell into these holes and could not escape, but sadly none have yet been found on the island.

3.6 What is Chalk?

Chalk is a limestone, a sedimentary rock composed of calcium carbonate, usually in the form of calcite. The calcite of chalk is formed by the accumulation of microscopic phytoplankton on a deep-sea floor generally between depths of 100 – 600 metres. These phytoplankton are mostly single-celled algae with an internal, multi-element spherical 'shell' called a coccolithophore. Zooplankton, in the form of foraminifera, also with delicate calcite skeletons, make up some of the chalk sediment (*Fig. 3.25*). Limestone is an easy rock to identify, as it fizzes when you drop dilute hydrochloric acid or

Figure 3.25

Top: Coccoliths from soft chalk in the UK.

Right: Photo of a modern coccolithophore, *Emiliania huxleyi*, taken using a scanning electron microscope (SEM). Scale bar *Top* = 1 μm, *Right* = 2 μm.

Photos courtesy of Andy Gale.

vinegar onto it. Deposition of the chalk a long way from any land-derived sediment source results in a high-purity chalk, giving it its pure white colour. Chalk that looks grey has some clay mixed in, meaning it was deposited closer to shore than the pure white chalk.

The chalk is the reason islanders have issues with limescale. The rainwater filters through the chalk, dissolving up the calcium and magnesium minerals along the way. When the water evaporates, the calcium and magnesium salts are left behind, creating a limescale residue. Water on the island is some of the hardest water in the country as it comes from chalk aquifers.

3.7 Flint Formation

Flint is a cryptocrystalline variety of quartz, meaning the crystals of quartz are so minute, you struggle to see them even under a high-powered scanning electron microscope. Flint can vary in colour from white, black, and grey when fresh. Orange and red hues are common colours of weathered flints derived from the local Ice Age river gravels (*see page 98*). The surfaces of flints are glassy when broken, and freshly broken edges with conchoidal fractures are razor sharp. For this reason, prehistoric humans used flint for tool making over any other stone.

Never attempt to smash or break flint stones! If you find a fossil in a lump of flint, collect the whole block. Whilst breaking flints, small micro shards can easily fly into your eyes if you're not wearing safety googles, sharp edges can slice fingers open, and the fossils usually break with the flint.

How flint is formed is still a bit of a geological mystery. While some evidence has been found in deep-sea cores of the early flint formation process, the details of the chemistry are not completely understood. A leading hypothesis on its formation is as follows:

Flint is made of silica and is considered to have begun forming soon after the deposition of chalk. Organisms such as sponges, radiolaria and diatoms, use silica to make biogenic opal skeletons. When these organisms die, the biogenic opal is scattered on the seabed and becomes incorporated in the accumulating chalk sediment.

At shallow sediment depths, the biogenic opal dissolves, causing the pore water to be supersaturated in silica. At the same time, at sediment depths of >10 cm, decomposing organic material produces hydrogen sulphide. The hydrogen sulphide then migrates upwards through the sediment, meeting dissolved oxygen moving downwards. This oxidises the hydrogen sulphide to sulphate, with hydrogen gas as a by-product. The free hydrogen ions lower the PH significantly, turning the pore water acidic, which leads to the calcite chalk dissolving. The dissolving chalk leaves an increased concentration of carbonate ions, which triggers the silica to precipitate. The silica precipitates out as ultra-microscopic

Figure 3.26

Above: Flint infilled burrows at Alum Bay.

Right: A small bryozoan within a flint nodule at Freshwater Bay.

cristobalite spheres, which much later recrystallize into quartz, forming the hard flints we find today.

Flints often take on the shape of burrow systems (*Fig. 3.26*). This is due to the burrows having a higher organic content than the surrounding chalk causing the chemical conditions for flint formation to occur preferentially in or around burrows.

3.8. Chalk – Paleogene unconformity

There is a large unconformity (time gap) on the top of the chalk, with half of the Campanian and all of the Maastrichtian chalk missing, along with most of the Paleocene, creating a gap of about 500 metres of rock, which represents over 20 million years. This unconformity means the famous Cretaceous – Paleogene (K/Pg) boundary at 65.5 million years is absent. The K/Pg boundary is

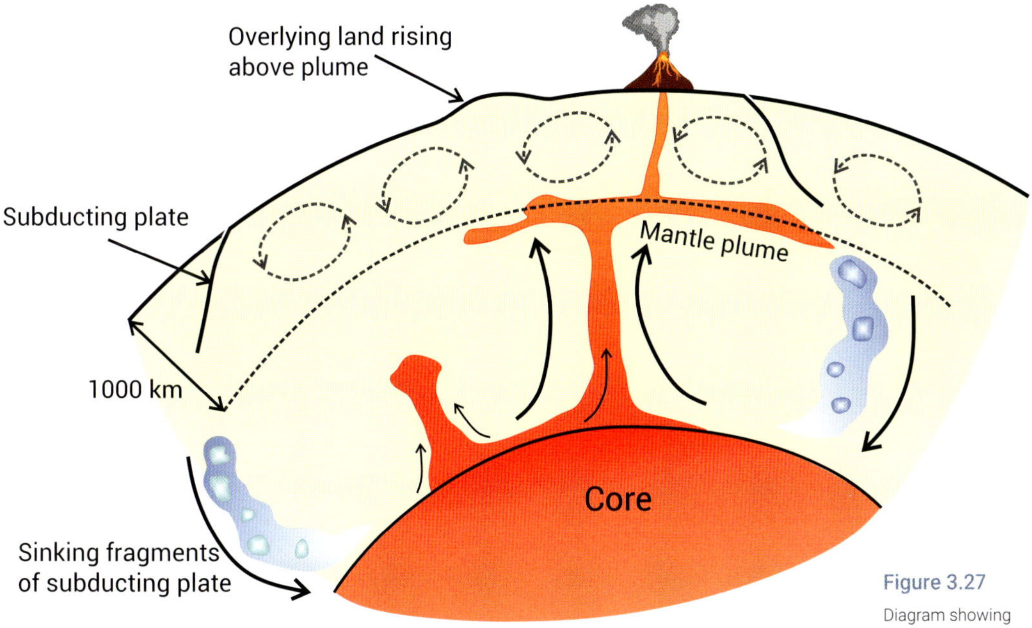

Figure 3.27
Diagram showing the formation of a mantle plume.

absent in the whole of the UK, with the only known boundaries being located in the North Sea and English Channel cores.

Likely, this unconformity was the result of an uplifting event, causing significant erosion to the upper portion of the chalk. This long-term erosion formed a peneplain (low-relief landscape) followed by the deposition of the Reading Formation.

The uplifting event has been attributed to the development of the Iceland mantle plume in the early Paleogene. Hot rock rising from the mantle (*Fig. 3.27*) beneath what is now Iceland starts to spread out into fingers once it reaches the asthenosphere (around 1000 km below the surface). One of these fingers reached Scotland, where it then branched out underneath Ireland, Wales, and England, with a large hotspot in the Irish Sea (*Fig. 3.28*).

This caused the overlying land to rise, and then subsequently be eroded. Researchers have been able to piece this story together by looking at the ages of the chalk across England and mapping these, which results in a 'bullseye' pattern across the plume, with the oldest chalk in the middle and the youngest chalk restricted to the margins (*Fig. 3.29*).

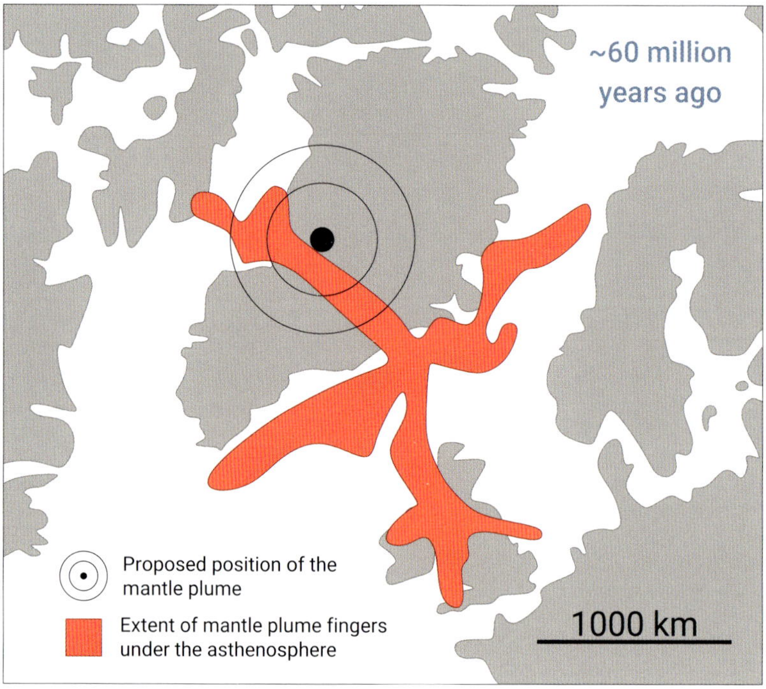

Figure 3.28

Diagram showing the extent of the Iceland mantle plume during the early Paleogene.

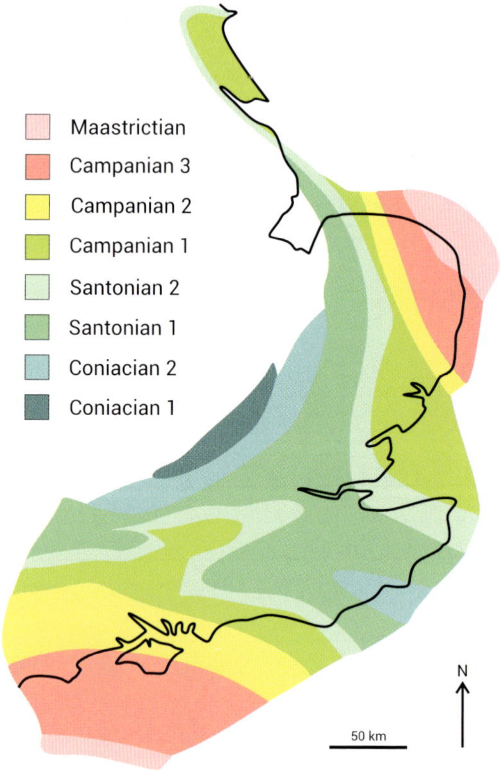

Figure 3.29

Diagram showing the age of the chalk across southern England, showing the bullseye pattern around the plume.

4. PALEOGENE GEOLOGY

Age: Paleocene - Oligocene (56 – 32 million years)

The Paleogene outcrop on the island covers the entirety of the northern portion of the island where it forms low rolling countryside. These sediments were deposited in a variety of different environments representing fluviatile, lacustrine estuarine and marine environments. There is a 20-million-year unconformity on the top of the chalk formations (see above). The Paleogene stratigraphic sequence on the Isle of Wight represents the most extensive and complete Paleogene succession in the UK and has been studied since the 1800's. Today, the sequence provides a globally important stratigraphic record and is a rich source of Paleogene fossils.

4.1 Lambeth Group

Age: Paleocene - Eocene, Thanetian - Ypresian (57 – 54 million years)

Reading Formation

The Reading Formation represents the oldest Paleogene strata in the Hampshire Basin dating to the late Paleocene or early Eocene (latest Thanetian - Ypresian). At Alum Bay, the formation is 26-35 metres thick and is between 35-40 metres thick at Whitecliff Bay. The Reading Formation unconformably overlies the Cretaceous White Chalk Subgroup (*Fig. 4.1*). The surface of the unconformity is characterised by large hollows up to 2 m in diameter, formed by subaerial dissolution processes forming sinkholes in the chalk prior to the deposition of the Reading Formation (*Fig. 4.2*). The base of the formation was deposited in a marine environment characterised by a bed of fine to medium grey sand with rounded flint stones, fresh nodular flint, and rounded blocks of chalk. These silts and sands are about 20 million years younger than the underlying chalk, indicating that a period of uplift and erosion preceded their

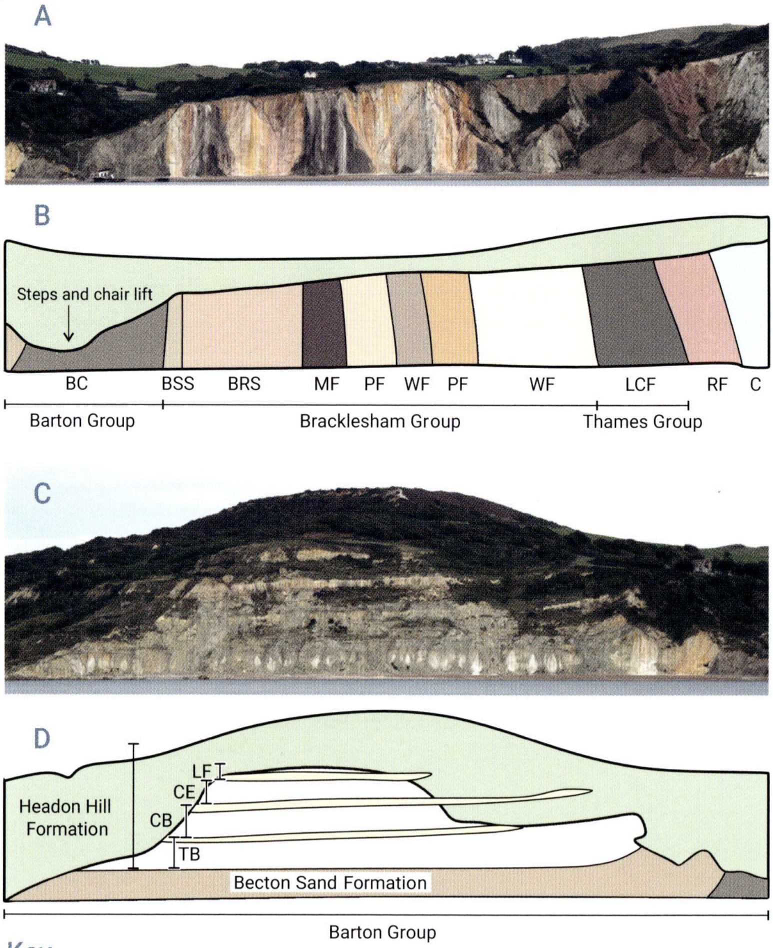

Figure 4.1

Stratigraphy of Alum Bay.

A, an offshore photograph of the southern portion of the bay, with B, an interpretive diagram. C, offshore photograph of Headon Hill with D, interpretive diagram.

Key

BC: Barton Clay Formation.
BRS: Branksome Formation.
BSS: Boscombe Formation.
C, Chalk.
CB: Colwell Bay Member.
CE: Cliff End Member.
LCF: London Clay Formation.
LF: Lacey's Farm Member.
MF: Marsh Farm Formation.
PF: Poole Formation.
RF: Reading Formation.
TB: Totland Bay Member.
WF, Wittering Formation.

deposition. The sands are overlain by mottled red and purple clays deposited on a river floodplain, and the colouration was caused by soil formation. At this time, much of southern England was a low-lying landmass of rivers, tropical forests and savannah, with a warm shallow sea to the east and uplands to the north and west.

The Reading Formation is usually poorly exposed, often obscured by mudslides or covered with vegetation. On the mainland, these deposits have produced fossils of crocodiles, early mammals and 'terror birds', but on the island, they are largely unfossiliferous.

Figure 4.2

Top: Reading Formation exposure at Whitecliff Bay, showing *Left:* mottled paleosols.

4.2 Thames Group

Age: Eocene, Ypresian (54 – 48 million years)

Harwich Formation

The Harwich Formation comprises several metres of glauconitic sands and silts, with a basal pebble bed containing rounded flints derived from the chalk and hard red nodules from the Reading Formation. At Whitecliff Bay, the formation is 2.3 metres thick and has a sharp contact with the underlying Reading Formation.

Figure 4.3

Exposure of the London Clay Formation at Whitecliff Bay, showing folded strata.

At Alum Bay, it is 4.3 metres thick, with a burrowed surface and is often referred to as the "London Clay basement bed". On the island, the London Clay is generally less fossiliferous, but fossils of molluscs and crustaceans are found.

Whitecliff Bay

London Clay Formation

At Alum Bay, the London Clay is 70-73 metres thick and is usually well exposed (*Fig. 4.3*), although sometimes the exposure can be concealed by mudflows. It comprises of alternations of blue to brown clays and buff sands. Well-preserved calcareous fossils occur in this formation, except in the upper units where only moulds are left due to leaching. The contact with the Reading Formation is characterised by a burrowed erosion surface, with some burrows preserved as ironstone casts.

The London Clay is divided into 5 divisions by King (1981), A-E, representing 4 marine transgressive-regressive cycles. The transgressions are marked by beds of rounded flint pebbles and glauconitic sands overlain by clays and silts deposited in deeper water under low-energy shelf conditions. The cycle is terminated by shallow marine sands formed during a sea-level fall.

The basement bed comprises 4.3 metres of glauconitic clays with shelly lenses. The succeeding subdivisions consist of 3 coarsening upwards units. Unit 1 (16 m) begins with silty and sandy clays, with common bivalves (*Pholadomya)* and serpulid worm tubes (*Ditrupa* sp.*)*. This passes up into silts, then into sandy silts. The bivalve *Panopea* in life position is abundant in the sandy silts, along with a diverse invertebrate fauna. Unit 2 (38 m) comprises silty clays, passing up into interbedded fine sands and clays. Shell seams and lenses can yield the bivalves *Pinna*, and *Arctica*, and the high-spired gastropod *Turritella*. Unit 3 (11 m) is composed of sandy silts.

4.3 Bracklesham Group

Age: Eocene, Ypresian - Lutetian (48 – 41.2 million years)

The Bracklesham Group is the most prominent part of Alum Bay, as the sand bodies within the succession create the famous 'coloured sands'. It is divided into four distinctive formations, the Wittering, Earnley, Marsh Farm and Selsey formations, all of which are fossiliferous. The succession is 220-230 metres thick, representing non-marine to marginal marine environments. In Alum Bay, the group is virtually unfossiliferous, except for a single horizon containing plant remains (the Alum Bay Leaf Bed), but in Whitecliff Bay abundant marine fossils are present throughout the succession.

The Wittering Formation

The Wittering Formation is the lowest unit of the Bracklesham Group, where it is 53 metres thick, whereas the exposure at Alum Bay is 43 metres thick. Within the upper part of the formation at Whitecliff Bay is a 1 metre thick lignite bed (coal) overlying a paleosol with abundant vertical rootlets.

The Wittering Formation consists of bioturbated to laminated clays and sands. A wide variety of sedimentary structures are to be found, including lenticular and flaser bedding. The sands that appear in this unit may be cross-bedded. There is a shell bed full of the carditid bivalve *Venericor*, 7 metres from the base of formation.

The higher part of the Wittering Formation (23 m) is composed of sandy muds within situ rootlets. This unit is predominantly heterolithic, with a rootlet bed also occurring at the top of the unit. In Alum Bay, it is composed mainly of 25 metres of sands with cross bedding. Also, present in the lower part of the unit are pebbly lenses and thin beds with *Ophiomorpha* burrows.

Poole Formation (west only)

In Alum Bay, this formation, approximately 45 metres thick, indigitates with the Wittering Formation. It is divided into two units, the so-called lower and upper beds. The lower part (23 m) is characterised by a fine to medium-grained sandstone, which fines upwards with both planar bedding and cross bedding present. It is overlain by a bioturbated fine-grained sandstone, with quartz pebbles and rounded flints. The upper part (25 m) is a fine to coarse-grained sandstone with cross-bedding. Its basal boundary is sharp, likely representing an erosion surface.

Earnley Formation (east only)

This formation is exposed in Whitecliff Bay, where it is 22 m thick. It was deposited in a marine, inner neritic environment, and comprises highly glauconitic, fine-grained sands that are well bioturbated. Its basal contact is a burrowed omission surface representing a transgressive horizon. This formation is rich in marine fossils, including several storm concentration beds, mostly yielding the bivalve *Venericor* and the gastropod *Haustator* (*Fig. 4.4*). As well as the giant foraminifera *Nummulites laevigatus* which is abundant in the upper 4.5 m.

Figure 4.4

Top: A storm accumulation shell bed comprised of *Vemericor* bivalves within the Earnley Formation.

Left: A storm accumulation shell bed comprised of *Haustator* gastropods within the Earnley Formation.

Marsh Farm Formation

The Marsh Farm Formation, 10 m thick, is a heterolithic sediment with laminated clays deposited in a marginal marine and inner neritic setting.

Branksome Sand Formation (west only)

The Branksome Sand Formation (65-70 m) consists of a lower and thicker unit of fine to medium grained sands with *Ophiomorpha* burrows and thin pebbly layers. This is overlain by a thinner unit of interbedded blocky massive clays with rooted lignite and silt to very fine sands.

Selsey Formation (east only)

The Selsey Formation is only seen on the shore at Whitecliff Bay and is up to 43 metres thick. It is deposited in a marine, inner neritic environment. It's sparsely glauconitic, and consists of shelly silts, clays and thin sands. Its basal contact is piped in burrows down into the underlying Marsh Farm Formation and is an omission surface. The so-called "*Cerithium giganteum* bed", a type of gastropod, is present at the base and a *Nummulites variolarius* bed is present in the uppermost few metres. The formation has been dated as mid to late Lutetian in age.

Boscombe Sand Formation (west only)

The highest unit of the Bracklesham Group is the Boscombe Sand Formation, which is 14 m thick. Composed of fine to medium grained sand, with a gradational base from the underlying Branksome Sand Formation. The formation coarsens upwards, with a thin bed of rounded and chatter marked flint stones.

The Alum Bay coloured sands.

About 20 colours are recognisable within the Alum Bay sands. Each reflects mixtures of 6 main colours: yellow, white, brown, green, red, and black (*Fig. 4.5*). The sands are predominantly quartz, with the colours only occurring as a result of staining or the presence of other materials mainly oxides of iron.

The white sands are composed entirely of quartz grains with no mineral staining. Black sands get their colour from small particles of lignite, derived from plant material. Glauconite grains, a Fe- potassium, alumina-silicate mineral, give some of the sands a greenish colour. Glauconite, as a somewhat unstable mineral, weathers when exposed to the atmosphere, decomposing to produce a selection of iron oxides. These oxides give various shades of yellows, browns, and reds; Hematite also gives a red colour, while limonite produces yellows and browns.

Figure 4.5

Outcrop of the Bracklesham Group at Alum Bay showing the famous coloured sands.

4.4 Barton Group

Age: Eocene, Bartonian (41.2 – 37.7 million years)

Overlying the Bracklesham Group is the Barton Group, which includes the Barton Clay and Becton Sand formations. In total the group is around 118 metres thick. At Alum Bay, the boundary between the Barton Clay Formation and the Boscombe Sand Formation is marked by a flint pebble conglomerate.

Barton Clay Formation

The Barton Clay Formation is generally composed of interbedded, bioturbated silts and clays, and is highly glauconitic at some levels. It is 83 metres thick at Alum Bay and at Whitecliff Bay, it is 45 metres thick and mostly comprises bluish-grey silty clays. There are two beds in the middle of the formation containing *Nummulites*. It was likely deposited in an inner to mid-coastal environment.

The Becton Sand Formation

The Becton Sand Formation consists predominantly of sandy silts with glauconitic sand. It is fossiliferous, with the bivalve *Chama* being particularly common. At Alum Bay, the exposure is 7 metres thick, where it is comprised of silty clays and blue-grey, and the

gastropod *Turritella* sp. and bivalve *Chama squamosa* are abundant at the base of the formation. At Whitecliff Bay, the formation is almost twice as thick at 15 metres, consisting of highly bioturbated muddy sands. However, it is poorly exposed (*Fig. 4.6*).

Figure 4.6

Exposure of the Becton Sand Formation at Whitecliff Bay.

4.5. Solent Group

Age: Eocene - Oligocene, Priabonian - Rupelian (37.7 – 32 million years)

Overlying the mid-Eocene Barton Group is the Solent Group. At up to 190 m thick it consists of 3 distinct formations, the Headon Hill,

Bembridge Limestone and the Bouldnor formations. Lithologically, the group is dominated by clays, silts, sands, and freshwater limestones. It is the youngest extant Paleogene strata preserved in the Hampshire Basin, spanning the onset of the late Eocene 37.8 million years ago to the early Oligocene ~32 million years ago.

The Solent Group saw a major change in the depositional environment of the Hampshire Basin, representing mainly freshwater and brackish deposition on a low-lying coastal plain, as opposed to the marine environments of the earlier Eocene. The landscape would have been similar to the subtropical Florida Everglades with wetlands, floodplains and lakes with drier areas of sequoia, cypress and palm woodlands, and a coastal fringe of lagoons and estuaries. The sea would still occasionally flood over the area briefly, depositing brackish and in one case, the Brockenhurst Bed in Whitecliff Bay being fully marine. As such, it yields an important assemblage of terrestrial fossils at this time, particularly mammals and plants, with insects also known to occur at some horizons.

The Solent Group is exposed along the entirety of the northern coastline of the Isle of Wight from Alum Bay in the west to Whitecliff Bay in the east and dominating the island's Paleogene exposures. The lowest part of the group, the Headon Hill Formation underlies much of the New Forest, but the higher Bembridge Limestone and Bouldnor formations are exclusive to the Isle of Wight.

Headon Hill Formation

The Headon Hill Formation is the lowest of the 3 formations of the Solent Group, taking its name after Headon Hill near Totland on the Isle of Wight. Here, on the cliffs of Headon Hill the full western sequence of the formation is exposed (*Fig. 4.1*). The formation dates to the late Eocene and consists of 8 members: the Totland Bay Member, Colwell Bay Member, Linstone Chine Member, Hatherwood Limestone Member, Cliff End Member, Nettlestone Member, Lacey's Farm Limestone Member, Fishbourne Member. The formation is composed of clays, sands, sandstones, and several freshwater limestones, representing a marine regression.

The Headon Hill Formation succession is complex and displays significant lateral variation across the northern Isle of Wight, with the sequence seen in the west very different to that seen in the east. The Seagrove Bay and Nettlestone members are notable examples. These members are absent in the western Isle of Wight, whereas the others like the Hatherwood Limestone and Linstone Chine members are absent in the east. Even those members that extend across the island such as the Totland Bay and Osborne Marls members, show lateral variation in both lithology and thickness.

The Headon Hill Formation is the only formation of the Solent Group present on the mainland, being distributed across both the northern Isle of Wight and the New Forest, with a famous fossiliferous exposure of the Totland Bay Member at Hordle, on the Hampshire coast.

The Totland Bay Member is the lowest member of the Solent Group, overlying the Becton Sand Formation of the Barton Group beneath. It consists of fine-grained mottled pedogenic clays, sands, thin lignite beds and several beds of limestone, deposited in a coastal floodplain environment, representing alluvial overbank areas, back-swamps, river-mouth sandbars, river channels and floodplain lakes. Common fossils include the gastropods *Viviparus*, *Galba* and *Planorbis*, and the freshwater bivalve *Potamomya*.

The Colwell Bay Member overlies the predominantly freshwater deposits of the Totland Bay Member and consists of sands and mud, which are often glauconitic. It is thought to represent fully marine to estuarine deposition in a large marine embayment fringed with estuaries and coastal lagoons, with salinity fluctuations throughout the member, especially in its upper part. This member is highly fossiliferous and yields diverse marine-brackish molluscan assemblages including the bivalves *Streostrea* and *Sinodia* and gastropods such as *Editharus*, *Pictoneritina* and *Potamides*.

The Linstone Chine Member is a variable and localised member of the Headon Hill Formation primarily composed of buff-coloured fine-grained sands overlying the Colwell Bay Member, with a maximum thickness of 5.2 metres. It is laterally discontinuous and best developed at Linstone Chine in Colwell Bay. It is absent from the eastern Isle of Wight. It was likely deposited in a shoreline/river-mouth sandbar setting. Common fossils include the brackish gastropod *Batillaria*.

The Hatherwood Limestone Member is an approximately 9 m thick limestone with an outcrop restricted to the very western Isle of Wight and best seen on the cliff of Headon Hill. It comprises several limestone units of subtly varying lithology, with limestone conglomerates and an intercalated lignite bed. As a result of this, the depositional environment is complex. The dominant limestones are thought to represent deposition in a very shallow ephemeral lake environment, like other limestones in the Solent Group. The lignite bed most likely represents an influx of organic-rich mud composed of fine plant debris deposited through overbank flooding. The conglomerates are evidence of contemporaneous erosion.

The presence of foraminifera at one horizon, however, indicates an intermittent lagoonal connection. Fossils of the freshwater pulmonate snails *Galba* and *Planorbis* are common in the limestone, along with the gyrogonites of charophytes (a type of green algae).

The Cliff End Member overlies the Hatherwood Limestone at Headon Hill, the Linstone Chine Member in the northwestern Isle of Wight, and the Colwell Bay Member in the eastern Isle of Wight. It is around 20 m thick and consists of two distinct lithologies, fine shaly laminated clays overlain by mottled pedogenic clays. Shell beds containing the bivalve *Potamomya* and the gastropod *Viviparus* are common. The member represents initial brackish lagoonal deposition followed by overbank floodplain deposition. The distinct nature of the two lithologies that comprise the Cliff End Member led King (2016) to propose their division into the Cliff End Member, comprising solely the lower shaly strata and the Fort Albert Member formed from the upper mottled clays.

The Nettlestone Member is limited to the area around Seaview and St Helens, where it overlies the mottled pedogenic clays of the Cliff End Member. The Nettlestone Member consists of fine sands, sandstones, and limestones with shell fragments, deposited in a river channel.

The Lacey's Farm Limestone Member is a continuous though variable member of the Headon Hill Formation cropping out at Headon Hill, Fort Victoria, Binstead, Seaview and Whitecliff Bay. The member is named after a now non-existent exposure at Lacey's Farm Quarry in Totland. At Headon Hill, Lacey's Farm Quarry and throughout much of the western Isle of Wight, the member consists of limestone and greenish calcareous sandstones and sandy limestones. In the eastern Isle of Wight, green marly clays dominate the member with some thin limestones.

The Fishbourne Member overlies the Lacey's Farm Limestone Member and is found across the northern Isle of Wight, with the principal exposures at Fishbourne, Binstead, Seaview and Whitecliff Bay. It consists of shaly laminated clays deposited in a brackish lagoon. Within this member is the Chapelcorner Fish Bed, a Konservat-Lagerstätten, yielding exceptionally well-preserved, fully articulated fish and crustaceans (*see page 236*).

The Osborne Marls Member is found across the northern Isle of Wight and mainly comprises mottled pedogenic clays, very thin lenticular beds of fine sand and lenticular shell beds in the upper sections totalling ~20 m thick in its entirety. Much of the member has been subject to intense soil forming processes and is strikingly mottled with bright reds, pinks and greens, and as such is easy to recognise. The member is considered to represent overbank deposition on a seasonally inundated floodplain. The mottled clays exhibit pedogenic features including mottling, desiccation cracks and rootlets.

The Seagrove Bay Member is restricted to the northeastern Isle of Wight, exposed at Binstead, Seaview, St Helens and Whitecliff Bay. Its stratigraphic position correlates with the upper sections of the Osborne Marls Member seen in the western Isle of Wight, and thus in the east it directly underlies the Bembridge Limestone Formation as the uppermost member of the Headon Hill Formation. It consists of fine to coarse sands and sandstones with commuted shell fragments and lags of flint pebbles, deposited in a fluvial channel.

Figure 4.7

Right: A, laminated clays and shell beds of the Hamstead Member at Bouldnor.

B, foreshore exposure of the Nematura bed (Hamstead Member) at Bouldnor, showing the bivalve *Polymedsoda convexa* preserved with original colour banding.

C, the outcrop of the Bembridge Limestone Formation at Whitecliff Bay.

Bembridge Limestone Formation

The Bembridge Limestone Formation is a sequence of fine-grained lacustrine limestones and clays deposited during the Late Eocene in a network of very shallow and likely ephemeral carbonate lakes, detached from the main floodplain river system (*Fig. 4.7*). The formation is not formally subdivided but typically comprises an upper and lower limestone unit separated by mottled lacustrine clays referred to as the 'middle muds', within which also lies a horizon of brackish origin containing the bivalve *Corbicula*. The gastropods *Galba* and *Planorbis* are common. Evaporite minerals including gypsum pseudomorphs are found in the lower limestone unit, which correlate with the Gypses de Montmartre of the Paris Basin.

Bouldnor Formation

The Bouldnor Formation is the uppermost formation of the Solent Group, overlying the Bembridge Limestone. It is divided into 3 members (Gurnard, Hamstead and Cranmore members) consisting of clays, muds, sands, and thin limestones, deposited in floodplain, lake, estuarine, lagoonal and in the case of the Cranmore Member, marine environments. It is the youngest stratigraphic unit preserved in the Hampshire Basin, spanning the very latest Eocene into the early Oligocene. The Eocene-Oligocene boundary likely occurs somewhere in the lower Hamstead Member, although the exact position of the boundary is still a topic of research and debate.

The Gurnard Member comprises a series of fine clays, thin limestones, sands, and shell beds deposited in a variety of depositional environments. The lower sections immediately overlying the Bembridge Limestone represent deposition in an outer estuarine environment, with thin laminated muds and shell beds containing marine-brackish molluscs. The Insect Limestone, a persistent bed of concretionary micritic limestone yielding insect fossils, represents sedimentation in a series of shallow hypersaline pools before a reversion to estuarine/lagoonal conditions (*see page 250* for more on the Insect Limestone). The upper sections comprise clays and mottled paleosols indicative of deposition in floodplain and lake environments.

The Hamstead Member comprises fine clays, sands and shell beds deposited in estuarine, lagoonal, floodplain and lake environments. Much of the sequence consists of mottled clays and thin sand lenses deposited on an overbank floodplain. Notable exceptions to the general freshwater deposition include the *Nematura* Bed (*Fig. 4.7*) and the White Band; horizons of lagoonal/ estuarine origin containing brackish molluscs, and acting as important marker beds in the sequence. The mammalian record of the Hamstead Member preserves the early Oligocene faunal turnover event of the Grande Coupure (*see page 244*).

The Cranmore Member is 9 metres thick and is the highest part of the Bouldnor Formation overlying the Hamstead Member. It is formed of dark clays with shell beds and nodule bands containing fossil faunas of near marine to marine origin, representing deposition in an outer estuarine to a shallow marine setting.

5. QUATERNARY GEOLOGY

Age: 2.5 million years – 8,000 years

The quaternary deposits on the island encompass both the Pleistocene and Holocene epochs. The layers of gravel often seen at the top of the cliffs around the island are the remains of riverbeds, deposited at the end of the last interglacial. The island at the time was most likely covered in patches of oak and hazel tree woodland, along with other varieties of native trees (*Fig. 5.1*).

Fossils reveal a fauna dominated by herds of bison, aurochs, and deer, with rarer remains of elephants, rhinos, and hippos. Based on similar aged remains around the country, it is also likely that the island was home to lions and hyenas. There is also evidence of human (*Homo sapiens* and *Homo neanderthalensis*) habitation in the form of a variety of flint tools, but no skeletal remains of humans from this period have been found. There are three key sites, Priory Bay, Great Pan Farm and Bleak Down which are nationally important due to the abundance of Palaeolithic artefacts found indicating lengthy periods of human occupation.

The geology of the Quaternary deposits comprises a patchwork of gravels, sands and clays across the island, with the correlation of beds highly questioned and very few reliable dates obtained. This is mainly due to the highly complex and laterally changing landscape of the Pleistocene to Holocene on the island. As a consequence, almost every outcrop has a different member name, within a broader formation encompassing all outcrops of a certain area.

As no one wants a blow-by-blow account of every patch of gravel and sand on the island, discussed below is a selection of the more interesting outcrops that are easily accessible.

Figure 5.1

Right top: Holocene gravels with in situ log at Brook.

Right: Windblown sands along Atherfield cliffs which now form a rabbit warren.

5.1 Clay-with-flints

The clay-with-flints are the oldest known Quaternary deposits on the island, with the flints derived from the dissolution of the chalk.

These deposits are reddish brown clay with lots of broken flints and partially dissolved chalk fragments, often infilling solution hollows. The clay-with-flint deposits range from Palaeogene to Pleistocene in age and can be best seen infilling solution hollows within the chalk at Alum Bay (see Figure 3.24). It's thought that much of the clay-with-flints deposits have been eroded during the mid-Miocene to Pleistocene.

5.2 Terrace Gravels

The Terrace gravels are fluvial aggradation of the Solent River and its tributaries. In simple terms, these are the sediments deposited within the river systems when the amount of sediment entering the river is more than the river can transport. On the island, the terrace gravels comprise 4 formations- Western Yar River Formation, Solent River Formation, Medina River Formation, and the Eastern Yar River Formation. These formations range from 3 – 5 metres thick, with increased thickness when associated with channels.

5.3 Western Yar River Formation

The Western Yar River Formation outcrops around the western Yar River and the southwestern coastline, consisting of a mixture of gravels, clays, and sands in association with the ancient Western Yar and its tributaries (*Fig. 5.2*). There are 7 members in the formation: Kings Manor Gravel Member, Downton Farm Gravel Member, Backet's Copse Gravel Member, Sudmoor Point Gravel Member, Freshwater Gravel Member, Headon Warren Sand, and the Gravel Member. Dates for the formation are not available, however woolly mammoth remains have been found in the Kings Manor Gravel Member, indicating a cold climate, and a suggested Late Devensian (31,000 – 16,000 years) in age. The Late Devensian saw an advancement in glaciation, the so-called Last Glacial Maximum, with glaciers reaching Norfolk and causing a change from a warmer climate fauna to a cool climate fauna (elephants to mammoths).

The Headon Warren Sand and Gravel Member is an isolated deposit across Headon Down and it represents a mix of an

Figure 5.2

Right top: Exposure of Quaternary fluvial sands and gravels at Freshwater Bay, resting upon *Right:* The Bembridge Raised Beach member.

alluvial fan of material washed down from the highlands south of Headon Down as well as some fluvial depositions.

5.4 Solent River Formation

The Solent River Formation was deposited within the lower reaches of the River Solent. The formation consists of 8 members: Bembridge Raised Beach Member, Wotton Gravel Complex Member, Steyne Wood Clay Member, Bouldnor Copse Gravel Member, Knight's Cross Gravel Member, Twenty Acre Gravel Member, Dunnage Copse Gravel Member, and the Great Wood Gravel Member. The members have been dated between 367,000 and 115,000 years old.

The Bembridge Raised Beach Member is one of the more impressive outcrops of the formation, exposed from Bembridge Point to Howgate Bay (*Fig. 5.2*). It is composed of two units; the lower unit made of flint gravel and sand lenses and is thought to represent a high energy beach and possibly a spit. The upper unit is clay, silt and sand with rounded fine flint gravel. It is thought to be a periglacial solifluction deposit. This is where the ground thaws after a period of freezing, and the soils become saturated in water. This then causes them to slide slowly downslope.

The Steyne Wood Clay Member outcrops at several localities across the north of the island, however, the best exposures are located at Priory Bay. The deposits at Priory Bay have been long associated with hominin occupation, with thousands of flint artefacts, including hand axes, having been found in the bay since 1886. This makes it the richest Palaeolithic site on the island, and one of the richest in the country.

5.5 Medina River Formation

The Medina River Formation occurs within the Medina River valley, composed mostly of gravel and clays. The formation consists of 7 members: Merstone Clay Member, Seaclose Park Gravel Member, Bridge Farm Gravel Member, Blackwater Hollow Gravel Member, Robins Hill Paleosol Member, Bleak Down Gravel Member and the St. Georges Down Gravel Member.

The Seaclose Park Gravel is thought to be associated with the Great Pan Farm site, an area of terrace gravel aggregations. The Great Pan Farm site is famed for its *bout coupé* hand axes made by Neanderthals. These are a rounded triangular axe only found in Britain and are between 59,000 and 41,000 years old. As well as the axes, lots of flakes, scrapers, and blades commonly occur.

5.6 Eastern Yar River Formation

The Eastern Yar River Formation are terrace gravels, considered to be the remains of a large Proto Eastern Yar river system, with the flints derived from the southern downs. The formation is split into two units. The first unit consists of 3 members: Cockerel Gravel Member, Hale Common Gravel Member, and the Langbridge Gravel Member, representing the main river channel. The upper unit encompasses the Bobberstone Farm Gravel Member, Whiteley Bank Gravel Member and the Froghill Farm Gravel Member which are associated with tributary streams of Wroxall Brook. How Carpenters Farm Gravel Member correlates with other deposits is currently unknown.

6. FOSSIL HUNTING

6.1 Safety and Responsible Collecting

Before heading out to collect, be sure you are prepared for the adventure ahead. Always be aware of the tides, and the weather and follow the collector's code.

Weather

- Check the weather and sea conditions. This is important for the south coast of the island, between Compton and St. Catherine's Point. Often rough seas can pick up large rocks and powerful waves wash up to the cliffs. Do not attempt to head out in these conditions. Too many fossil collectors have almost been washed out to sea attempting to collect in these conditions, no fossil is worth your life.

Cliffs and Foreshore

- Always look up, before you look down. Before approaching the cliffs, look at them. Check for any overhangs and cracks. Blocks on the beach with fresh grass, or with lots of fine debris indicate an active fall within the last high tide (*Fig. 6.1*). If these are present, DO NOT APPROACH THE CLIFFS!
- The cliffs are unstable all year, and rocks can fall at any time, so do not climb the cliffs, hammer into them, or sit too close.
- Hard hats should always be worn when you are close to the steeper cliffs. Even a small piece of rock falling from a height can cause serious injury.
- High Viz jackets/ vests are also sensible, especially when visiting isolated localities. This will allow you to be easily spotted by rescue groups if you get into trouble.

Figure 6.1

Unstable cliffs.

Left: A cliff fall at Compton Bay. Note the base untouched by the tide indicating this fall happened in the last couple of hours.

Above: A large mudslide at Brook partially covering a dinosaur foot cast.

- Avoid rockfalls and mudflows, especially during winter months when these are more active. It is all too easy to get stuck in soft mud! In the wetter months, mudflows can be rapid (*Fig. 6.2*).
- Rocks covered by the high tide can be slippery, covered with fine silt or green algae, so take care when walking over these surfaces.
- Many of the islands' cliffs and foreshores are protected places and some are sites of special scientific interest (SSSI's). Permission is needed for serious investigations and excavations of fossils at these sites. Most fossils discovered on the island are found loose on the foreshores, just waiting to be picked up.

Figure 6.2

Landslides at Atherfield Point form a cutoff point at high tide.

Access

- Sites of Special Scientific Interest (SSSIs) protect the geology and wildlife. This means digging and hammering into the cliffs and foreshore and disturbing wildlife is prohibited. Refrain from causing damage to the coastline and keep a good distance from any wildlife you may encounter.
- Some of the island's beaches are privately owned, and there is no automatic public right of way, even in the intertidal region. Be respectful of this, making sure there is a public right of way where you would like to fossil hunt. Private beaches can be legally accessed by contacting the landowner for permission beforehand.
- Many of the island's cliffs are privately owned, especially along the southwest coastline and you require the landowner's permission to collect on this land.

Tides

Figure 6.3

Left: A lucky escape! Wellies stuck in deep mud at Brook roughs.

- It's always best to head onto the beach on a falling tide, or at low tide. But keep an eye on the tide once it has turned so that you don't become stranded or cut off (*Fig. 6.3*). Tide times can be found on a variety of websites and phone apps.
- On the island, it is best to use the 2-metre mark on tide curves. Anything under 2 metres usually means safe beach access depending on weather conditions. Always observe the beach before heading down, as stormy periods can lead to sets of large waves coming in occasionally. This can make incredibly dangerous conditions on the more remote parts of the island.
- Let someone know when you're heading out and planning to be back, especially when visiting more remote locations.

Collecting code of conduct

- If you find anything looking interesting in the cliffs or on the foreshore clays, do not try and remove it, take photos of the fossil, including some that have the cliffs in the background (for easy relocation), and report the findings. This way it can be thoroughly investigated, recorded, and scientifically collected (*Fig. 6.4*).
- Avoid over-collecting of fossils. Fossils are a finite resource and it's always best to take just a few representative examples of each fossil.

Figure 6.4

Photo of a large ammonite impression with scale in the field.

- When using hammers and chisels, be sure to wear appropriate PPE (goggles, gloves).
- Never hammer flint, as it shatters into small, extremely sharp shards, which will cut your hand open or penetrate your eyes all too easily. These shards also pose a real hazard to dogs and people walking barefoot on the beach.
- The Hanover Point dinosaur foot casts and footprints are protected. Leave the dinosaur footprints and foot casts on the beach, undamaged, for everyone to enjoy.

DO NOT ATTEMPT TO COLLECT OR HAMMER THEM.

Who to contact?

The coastal land around the island has multiple landowners, sometimes with the foreshore and cliffs having different ownership (*see Figure 6.5*). It is widely accepted that finds on the foreshore are ok to pick up and take home (except the dinosaur foot casts). However, if you find anything within the cliffs or foreshores, these need reporting to the relevant authorities or landowners. It is not

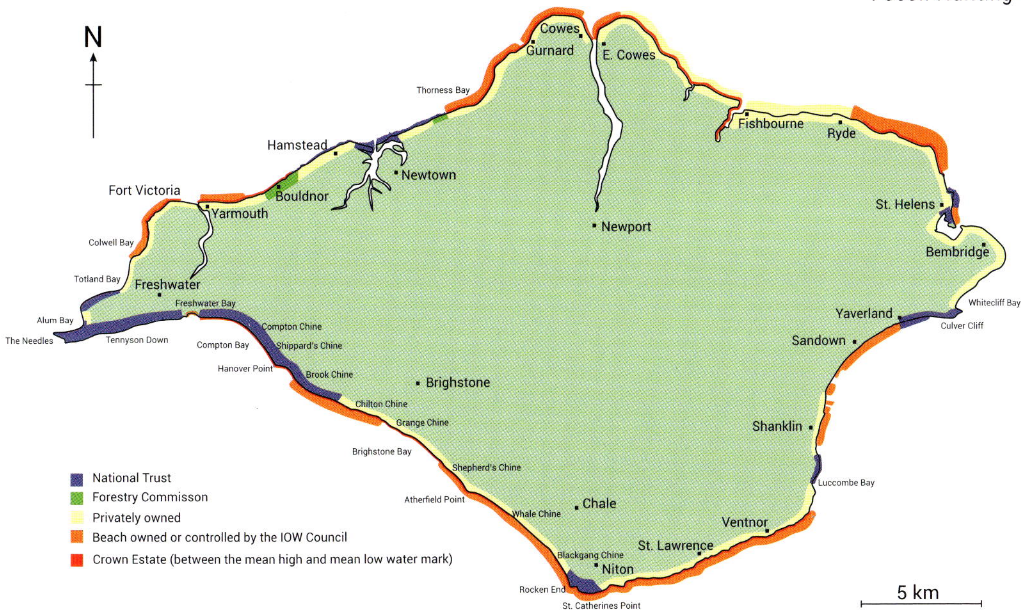

Figure 6.5

Map of coastal land ownership and management (as of 2024). For much of the coastline, ownership of the foreshore and cliffs can be different.

uncommon for rare fossils or complete skeletons to turn up, and these will need excavating by experienced teams.

For any in situ finds on National Trust land, contact the Isle of Wight team using the details below. The National Trust rangers will access the find and contact the relevant authorities if further investigation/excavation is needed.

National Trust Isle of Wight, Longstone Farmhouse,
Strawberry Lane, Mottistone, Isle of Wight PO30 4EA
Email: isleofwight@nationaltrust.org.uk
Telephone: 01983 741020

For all other in situ finds contact Dinosaur Isle Museum at Sandown, information is below:

Telephone: 01983 404344
Email: dinosaur@iow.gov.uk

6.2. Fossil Care

Some fossils are very delicate and fragile. Therefore, looking after your fossil finds, from the moment you find them through

Figure 6.6

A fossil lobster collected by Scott Cade, wrapped in tin foil to protect it and hold the fossil together for glueing and preparation at home.

to getting them home safely is especially important. These steps help to prevent damaging your precious fossils and increase their survivability whilst in transit.

Out collecting

Care should be taken when extracting fossils from the rock, and it is much better to remove the fossil in a block so that the fossil can be fully extracted at home under ideal conditions using the correct tools. For more fragile fossils, wrapping fossils individually in newspaper, kitchen or toilet roll, tinfoil or bubble wrap helps protect the fossils whilst they are rattling around in your backpack (*Fig. 6.6*). Tins and small plastic tubs recycled from takeaway meals are great for transporting delicate fossils. Cotton wool is not advisable as it leaves fine fibres on your fossils.

Recording your finds

Perhaps less than 0.1% of everything that's ever lived has become a fossil! Such rarity makes all fossils unique and potentially a valuable scientific resource recording the earth's ancient biological history. Therefore, it is important to record your finds with at least some basic information, allowing them to be scientifically useful if they are ever donated to a museum. Therefore, it is good practice to take a notebook with you when you go out collecting to record your discoveries. Although of course, you can record your discoveries on your mobile phone too.

Some information to include with your finds, and to be added to the specimen's label when you build a collection is suggested here:

- Date of collection
- Site of collection (GPS coordinates, what3words), as specific as you can be using local landmarks, the nearest village, chines, etc.
- If possible, record the formation of origin from which the fossil came, and note if it was collected in situ — including the specific bed or horizon, or a description of that bed (eg. colour, sand or clay) — or if it was picked up from the beach. If you find a fossil in situ, take a photograph before you extract it, and don't forget to add some sort of scale. A coin will do if you do not have a rule, and if you are broke, a finger will suffice.
- Because you might collect many fossils in a day, it is a good idea to number your notes for each fossil you find and put a piece of paper with the fossil with the same number. That will help avoid any confusion later on, especially if you have been to more than one locality in a day.

Don't worry if you cannot identify the fossil, as this can be done later, although if you do think you know what it is, do record it. The information about where the fossil came from may help in the identification process later on.

At home

Soak all fossils from the beach in freshwater for several days, changing the water at least once a day, then dry the specimen slowly. This will remove all the salt from your specimens and will help preserve the fossil. Sometimes the salt can crystallise within the fossils, causing them to crack or fall apart.

This is especially important for fossils that contain pyrite (fool's gold), as the pyrite can easily decay and crumble apart. It's best to store pyritised fossils in very dry, cool conditions, perhaps including a small sachet of silica gel to absorb any moisture. A coat of diluted (10%) paraloid B-72, an acrylic resin dissolved in acetone, can be used as a protective barrier or consolidant for fragile fossils. A thin clear varnish or PVA glue will work just as well.

The fossil wood from the Wessex Formation is often incredibly unstable, crumbling apart almost straight away and there's little you can do to preserve it.

6.3. What is a fossil?

Fossils are the preserved remains or traces of ancient organisms. The word fossil comes from the Latin *fossus*, meaning 'having been dug up'.

They form when an animal or plant dies, and its remains become buried in sediment. For the highest level of preservation, burial needs to happen rapidly after death, and the chemical process that forms fossils must occur quickly.

After burial, many physical, chemical, and biological processes turn the dead remains into fossils, many of which involve highly complex chemical reactions. Below is a simplified discussion of the most common modes of fossilisation that occur on the island.

Types of Preservation

Cast/Mould

These form when organisms buried in the sediment decay or dissolve away. Calcite, the mineral shells are made of, dissolves away easily in slightly acidic conditions. This leaves a cavity with the exact imprint of the organism's original shape. When rocks are broken, you can often see the imprints of fossils, these are called moulds. When the cavity is filled with sediment or minerals, taking the shape of the original fossil, it forms a cast. Both the cast and the mould are fossils (*Fig. 6.7*).

Internal moulds of fossils are also found commonly, where the chambers of gastropods, bivalves and ammonites have been infilled with sediment or minerals before the original shell is dissolved away, leaving a steinkern. Infilled ammonite chambers are commonly referred to as cats' paws.

Permineralisation

After burial, mineral-rich water migrates through the sediment and into the buried remains. The water deposits minerals into microscopic spaces, which over time crystallize. This process can occur in incredibly small spaces, including the spaces between bone cells.

Replacement/recrystallisation

In this mode, the original bone, shell, or wood is completely replaced by a different mineral, but still retaining the original shape. An example of this is petrified or silicified wood (*Fig. 6.7*).

Figure 6.7

A, external and internal moulds of *Viviparus* gastropods from the Vectis Formation.

B, a cut and polished section of a sponge replaced by flint, with quartz crystals inside some cavities.

B, collection of Dinosaur Expeditions C.I.C. Scale equals 10 mm.

A

B

Compression

Occasionally, the burial process causes the remains to be physically compressed by extremely high pressures. However, this compression can lead to distortion of the fossil, as it goes from a 3D object to an almost 2D object. Compression fossils are often found by splitting rocks along a natural bedding plane, resulting in a slab and counter slab, with fossil remains appearing on each surface. These are called parts and counterparts. This mode of preservation is most common with plant remains, some shelly fossil beds and fish beds.

Carbonisation

Fossils preserved this way have been reduced primarily to the chemical element carbon, preserved as a thin black film on the rock surface. These are typical of plant and insect remains.

Types of Fossils

Fossil collectors come across many types of fossils, and not all are the animal or plant itself. Here is a brief explanation of the types of fossils you may encounter:

Body fossils

Fossils of the animal or plant itself, such as bones, teeth, feathers, and shells.

Trace fossils

These are the tracks, traces and trails animals left behind during their lives. They document activities and behaviours such as walking, burrowing, and boring. The fossils include footprints, trackways, burrows, and borings in other fossils.

Chemical fossils

Sometimes when you break open a rock, it is possible to smell petroleum or a strong sulphur-like smell. These are the decomposition products of ancient animals and plants after decay, such as oil or natural gas. They are often regarded as chemical fossils as they are the preserved chemical molecules or their breakdown products that are the signature of ancient life. They are of little importance to fossil collectors but are scientifically important as chemical biomarkers are the oldest traces of life on Earth.

Pseudofossils

Erosion, such as wind, water and chemical action over time can sculpt rocks into a vast range of shapes, sizes, and colours. Sometimes these can produce shapes that can be easily mistaken for fossils. On the island, flint is often brought into the museum as its shape often resembles bones.

Reworked fossils

Fossils that have been eroded from the rocks and redeposited into younger sediments are said to be reworked. For example, a 130-million-year-old dinosaur bone eroding out and being buried in the sand of a present-day beach would be classed as reworked. Reworked fossils are usually highly abraded, while the contemporary fossils with them may preserve delicate structures such as spines.

6.4. Collecting on the Isle of Wight

Where to look

Searching for fossils is easy on the island. Almost any beach with some pebbles will yield fossils simply by looking through shingle banks on the beach (*Fig. 6.8*). Amongst the shingle, the sea has already washed the fossils from the rocks and cliffs, so there is no need for a hammer and chisel or any digging in the cliffs.

The best areas on the beach are the shingle banks closest to the sea, as these have been washed over by the last tide, with new fossils revealed every day!

The top of the beach tends to be picked over frequently, especially during the summer months. It's also not advised to go near the base of the cliffs or climb the cliffs due to the unpredictability of rock falls all year round.

Take some time to look at the shingle banks before getting your eyes close to the ground. The constant lapping of the waves sorts the pebbles according to their size and density. You might notice zones of deep brown pebbles and even greenish-grey pebbles. The latter are zones of iron pyrite (so-called fool's gold) pebbles, and these often contain fossils; usually bits of plant debris, but bones and teeth also occur within the pyrite shingle.

However, not all fossils are found on the beach! Fossils have been curiosities for centuries and have even been incorporated into buildings and walls all around the island, the most common being ammonites. Keep a lookout in the walls of older buildings for these fossils as you travel around the island (*Fig. 6.9*).

Figure 6.8

Photograph showing the *Top:* shingle banks and *Right:* pyrite lags along the southwest coast.

Building stones can also contain fossils, especially those made from the Bembridge Limestone, the Upper Greensand and even the Chalk formations. Looking carefully over the surfaces, it is often possible to spot molluscs and a variety of trace fossils (*Fig. 6.9*).

Do not try to remove any fossils from buildings or walls. These are beautiful features, often incorporated decades ago and should be left for everyone to enjoy. And it might be part of someone's house!

Figure 6.9

A, ammonite in the wall of the Godshill Methodist Church.

B, ammonite impression in a wall in Bonchurch.

C, fossilised oyster in the St. Boniface church in Bonchurch.

D, Bembridge Limestone blocks in Brook Lifeboat Station.

What to look for

Along the southwest coast between Compton Bay and Atherfield Point and Yaverland beach from the Wildheart Trust to Red Cliff, the Wealden Group strata yield dinosaur bones. These are usually black or dark, chocolate brown with a white honeycomb pattern on the broken surfaces. But they can also occur encrusted with calcareous algae if they have been sitting offshore for a prolonged period and therefore can be quite difficult to spot. It is a good idea to familiarise yourself with the shapes of bones to be able to spot them when they are covered in algae.

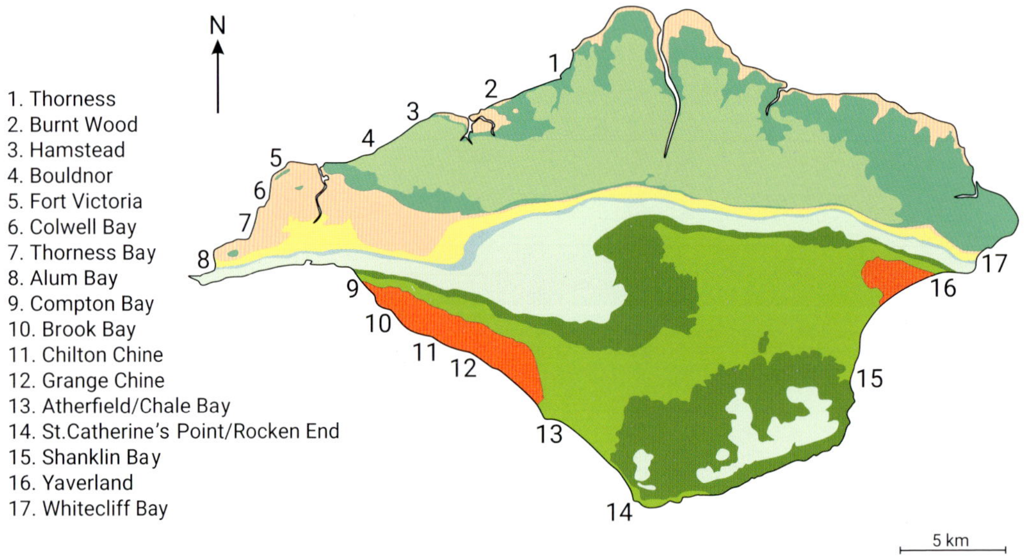

Figure 6.10

Map of the Isle of Wight showing popular fossil localities.

For other fossils, being familiar with the shapes, textures and colours of fossils helps significantly with spotting them in rocks and loose on the foreshore. If in doubt, look for stones with interesting shapes or patterns, or simply ones that stand out from the rest. These can be taken home to be identified at a later date or take a photo if you're unable to safely collect it.

6.5. Popular localities

1. Thorness

Thorness is a large bay that at first sight does not appear to be good fossil hunting territory. However, amongst the fine shingle of the upper part of the beach, at the high tide mark, lots of fossils can be found. The black and brown bones and teeth of fish, reptiles and mammals stand out against the white grit, making their discovery easy.

This locality is one of the most accessible sites on the island, a short, flat walk from the parking area at the back of the caravan park, suitable for all ages and mobilities (*Fig. 6.11*). This locality is particularly good for flint tools derived from the Pleistocene gravels.

Figure 6.11
Thorness Bay.

Geology. Bouldnor Formation (Gurnard Member), Bembridge Limestone Formation, Pleistocene river gravels.

Fossils. Eocene mammals, turtles, alligators, fishes, molluscs, insects, Cretaceous flint fossils, Pleistocene – Bronze Age flint tools, Ice Age mammal bones.

Access. Access to the bay is along Thorness Lane, located along Rolls Hill. Thorness Bay holiday park is halfway down Thorness Lane, to the right. Driving through the park access is via a dirt drive leading to the bay. There is parking at the top of the drive. Several public footpaths also lead to the bay, with the coastal path from Newtown taking you directly to the bay and can be followed along to Gurnard (*Fig. 6.12*). The closest bus route is the no. 32, which serves Pallance Road, and involves a minimum 30 minute walk to the bay via public footpaths.

2. Burnt Wood

Burnt Wood is famous for its abundance of Ice Age fossils, mostly washed in from offshore deposits, with common finds being deer and bison bones and teeth. Paleogene fossils, such as turtle shell fragments and alligator remains can be found loose on the foreshore,

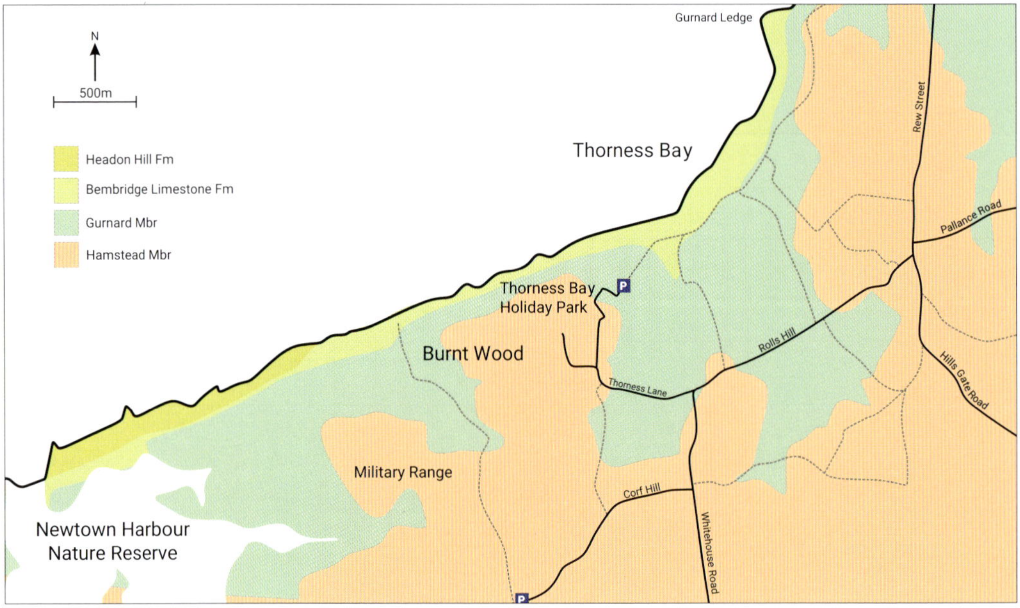

Figure 6.12

Map of Thorness and Burnt Wood

overall making for some great fossil hunting. However, due to the isolated nature of this locality, it is unsuitable for young children or those with mobility limitations.

It is important to note that this locality is an SSSI for wildlife and geology. So refrain from hammering into the cliffs or foreshore, and do not enter the Newtown Nature Reserve at the westernmost portion of the locality to avoid disturbing the wildlife. There are also many ground nesting birds, as well as seals, which live on the

Figure 6.13

Burnt Wood.

shingle banks at the entrance of the estuary, so maintain a good distance from these too. Most fossils are found along the first stretch of beach or out on the mudflats on an exceptionally low tide (*Fig. 6.13*).

Geology. Headon Hill Formation, Bouldnor Formation (Hamstead and Gurnard members), Bembridge Limestone Formation, Pleistocene river gravels.

Fossils. Eocene mammals, turtles, alligators, fishes, molluscs, Cretaceous flint fossils, Pleistocene – Bronze Age flint tools, Ice Age mammal bones.

Access. Parking is found along the main road and Locks Green Road, close to the Sportsman's Rest pub (postcode PO30 4PL). To access the beach, you must walk along Elmsworth Lane to Elmsworth Farm (there is no parking along this lane). A public footpath leads you through the farm, and alongside the copse to the beach. The walk from the main road to the beach is around 1.2 miles, taking around 30 minutes. The last 50 – 60 metres to the beach can be very marshy underfoot, so walking boots or wellies are essential, even in summer months. Be sure to stick to the public rights of way, as the western portion of land is a military firing range, and the eastern portion is a private farm and woodland. The foreshore can also be accessed via Thorness Bay, by walking west along the coast (*Fig. 6.12*). This locality is best accessed on spring low tides, with many Ice Age remains to be found on the mudflats.

3,4. Hamstead Hill and Bouldnor Cliff

Hamstead Hill and Bouldnor Cliff are famous localities for their high abundance of Eocene and Oligocene fossils. These are more challenging localities to visit, involving a long walk to access the coastline from either Cranmore Avenue or the viewpoint car park to the east of Yarmouth. The beach can be muddy in patches, and fallen trees make traversing the coastline difficult, so great care should be taken when visiting these beaches (*Fig. 6.14*). This site is unsuitable for young children or those with limited mobility.

Figure 6.14

Bouldnor looking east towards Hamstead.

Geology. Bouldnor (Hamstead and Gurnard members), Bembridge Limestone Formation, Pleistocene river gravels.

Fossils. Eocene mammals, turtles, alligators, fishes, molluscs, Cretaceous flint fossils, Pleistocene – Bronze Age flint tools, Ice Age mammal bones.

Access. The easiest, quickest, and safest way to access the coastline is from Bouldnor, where there is ample parking in the Bouldnor viewpoint car park, situated just outside Yarmouth on Bouldnor Road (A3054). There is a coastal path to the shore along Victoria Road that is reached with a short walk eastwards from the car park (*Fig. 6.15*). The walk from the car park to the shore takes around

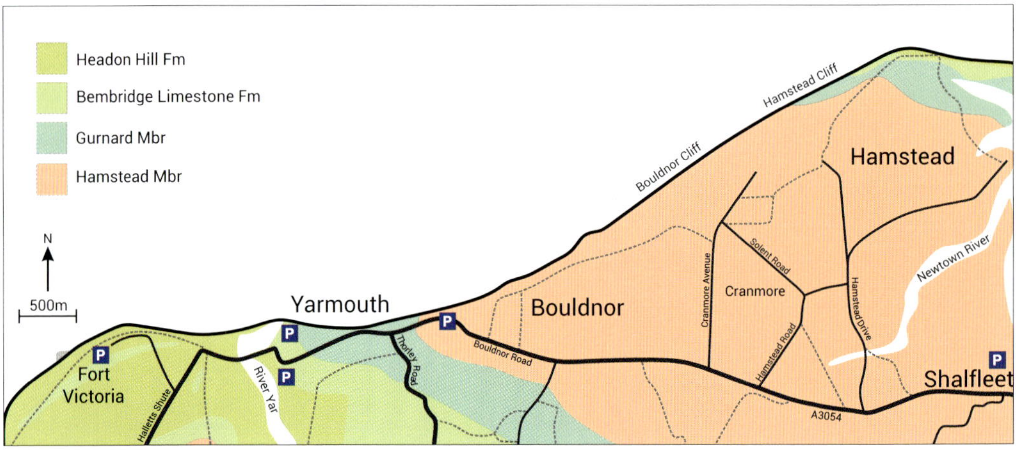

Figure 6.15

Map of Hamstead/ Bouldnor and Fort Victoria.

15-20 minutes. Limited parking for Bouldnor can also be found at the entrance to Bouldnor Forest. Several tracks and footpaths lead you to the foreshore but expect a 30-45 minute walk to reach the site from here.

Access to the rest of the coastline around Hamstead Ledge and Cranmore is difficult, requiring considerable walking and very limited parking. Parking for Hamstead can be found along Hamstead Drive, which, as with all the roads in this area, is private and permission must be obtained from the farm or Hamstead Grange. Space for 2-3 cars is available at the end of Cranmore Avenue, but again this is a private road, and any parking here should be done in a manner respectful of local residents. From here, the coastal path can be joined, heading east towards Hamstead and west towards Bouldnor. Parking for Hamstead is also available in the village of Shalfleet, with access via the footpath along Hamstead Drive, but with a walk of 50-60 minutes. You can access this location through public transport, the no.7 bus to and from Newport – Yarmouth whose route is along the A3054, with stops at each access point.

5. Fort Victoria

Fort Victoria was originally a 16th century defensive fort, having been rebuilt and demolished several times over the years. The remains have become the country park you see today. The archways now house many attractions, including a maritime museum, café, and toilet facilities. A walk along the roof of the fort provides superb views across The Solent. This is an ideal location for families of all ages to have a day out and go fossil hunting.

Fort Victoria yields a high diversity of Eocene fossils amongst the shingle and is very accessible. Fossiling here is best on a low or falling tide, with most fossils found below the high tide mark. Some sections of the beach further from the fort can be muddy and fallen trees make the western end of the beach difficult to traverse. However, the main portion of the beach is flat and easy going underfoot (*Fig. 6.16*).

Figure 6.16
Fort Victoria Bay.

Geology. Headon Hill and Bembridge Limestone formations.

Fossils. Eocene mammals, turtles, alligators, fishes, molluscs, Cretaceous flint fossils, Pleistocene – Bronze Age flint tools, Ice Age mammal bones.

Access. Fort Victoria a short distance west of Yarmouth along Hallett's Shute, at the end of Westhill Lane. It is only a thirty minute walk from Yarmouth town along the seawall. The no.7 bus route passes along the main road at the entrance of the park, with a 15 minute walk down the lane to the country park. Ample car parking is available. Access to the beach itself is easy, with a path leading directly onto the foreshore with no steps. There is also a coastal footpath from Colwell Bay which is around an hour's walk (*Fig. 6.15*).

6. Colwell Bay

Colwell Bay is a great locality for a wide range of fossils, from turtle shells, sharks' teeth, and molluscs. The fossiliferous section of the beach is located at the northern end, beyond the sea wall and sea defences. This section is more difficult to access due to having to walk over the sea defence blocks or climb over the groynes on a low tide. For this reason, it's not suitable for those with limited mobility (*Fig. 6.17*).

Fossils are found loose on the foreshore, or poking out the cliffs,

Figure 6.17
Colwell Bay.

needing only to be pulled out by hand; specialist tools (hammers, chisels) are not usually required. A small screwdriver or penknife is ideal for prising out the fossils sticking out of the clay. This locality is an SSSI, so refrain from digging into the cliff.

Geology. Headon Hill Formation.

Fossils. Eocene mammals, turtles, alligators, fishes, molluscs, Cretaceous flint fossils.

Access. Colwell Bay is easily accessible with roadside parking along the Colwell Chine Road (A3054) and a carpark near the seafront. The no.7 bus route from Yarmouth passes along Colwell Road, with just a 5–10 minute walk to the bay. Several footpaths lead to Colwell Bay, with a coastal footpath leading to Totland Bay as well as Fort Victoria (*Fig. 6.18*). The beach is easily accessible, with a seawall running along the top. There are several sets of steps along the seawall which allow access to the foreshore, however, the base of these can be incredibly slippery when covered in algae, so take care on these parts. It is advisable to visit this locality on a receding tide as the beach is inaccessible at high tide. To the north of the bay, the foreshore rocks are exposed and will be slippery, so care should be taken here.

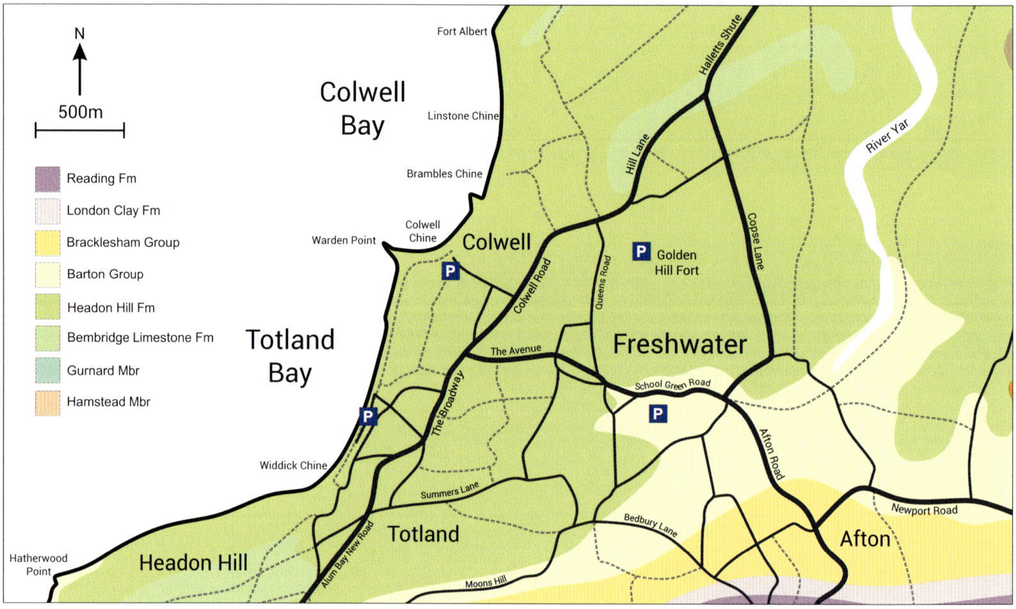

Figure 6.18

Map of Colwell and Totland bays.

7. Totland Bay

Totland Bay is an ideal location for an easy day of fossil hunting for the whole family with picturesque views overlooking the pier and Headon Hill, and flat sandy shores. Situated close to Totland village, you are close to local shops, as well as public toilets being situated at the centre of the bay. There is also a pub located to the east of the seawall. With all the fossil finds to be found on the foreshore, there is no need for special equipment, a small hammer at most to break open nodules (*Fig. 6.19*).

Figure 6.13

Totland Bay.

Geology. Headon Hill Formation.

Fossils. Eocene mammals, turtles, alligators, fishes, molluscs, Cretaceous flint fossils.

Access. Totland Bay is easy to access with a road leading directly onto the seafront as well as several footpaths, with the coastal path along the seawall from Colwell Bay. Parking is limited on the seafront, which is something to consider during the busy summer months (*Fig. 6.18*). More parking can be found on the roads above the bay. The no.7 and no.12 bus routes enter Totland village, where a 10 to 15 minute walk takes you to the beach. Several sets of steps along the seawall provide easy access to the foreshore, however, the base of these can be slippery if covered with algae.

8. Alum Bay

Alum Bay is famous for its unique cliffs of brightly coloured sand, but the clays yield a variety of important fossils. Some of these can be found amongst the shingle on the beach, but the best place for fossils is at the southern end of the bay, just before the chalk cliffs (*Fig. 6.20*). Here the cliff is composed of the London Clay Formation, which is rich in marine fossils. Much of this formation has slumped onto the foreshore and is exposed, with all the fossils protruding from the clay, where they can easily be 'plucked' from the sticky clay. The cliffs here are protected and privately owned. Hammering the cliffs or climbing them is prohibited, along with the collection of the coloured sands. One layer in the cliffs here yields beautifully preserved leaves, however it is rarely exposed or found as fallen blocks.

Geology. Chalk, Reading Formation, London Clay, Bracklesham Group, Barton Group and Headon Hill Formation.

Fossils. Eocene mammals, turtles, alligators, shark teeth, molluscs, crabs, lobsters, leaves, Cretaceous flint fossils.

Access. The needles park has a carpark at the end of the Alum Bay New Road (B3322), located at the top of the cliff. There is also

Figure 6.20

Alum Bay showing the coloured sands.

a National Trust car park, a 30 minute walk away, at the top of Highdown Lane off Moon Hill (*Fig. 6.21*). The no. 7 and 12 bus routes and island coaster buses stop at the entrance of Needles Park. A footpath with a series of steps leads down to the foreshore, which can be accessed from the top of Alum Bay Chine or through Needles Park. A seasonally operated, weather dependent chair lift is run by Needles Park allowing for easier access to the foreshore. A mid to low tide is advisable.

9, 10. Compton/Brook

Compton and Brook bays are famous for the large dinosaur foot casts at Hanover Point, mostly attributed to iguanodontid

Figure 6.21

Map of Alum Bay and Headon Hill.

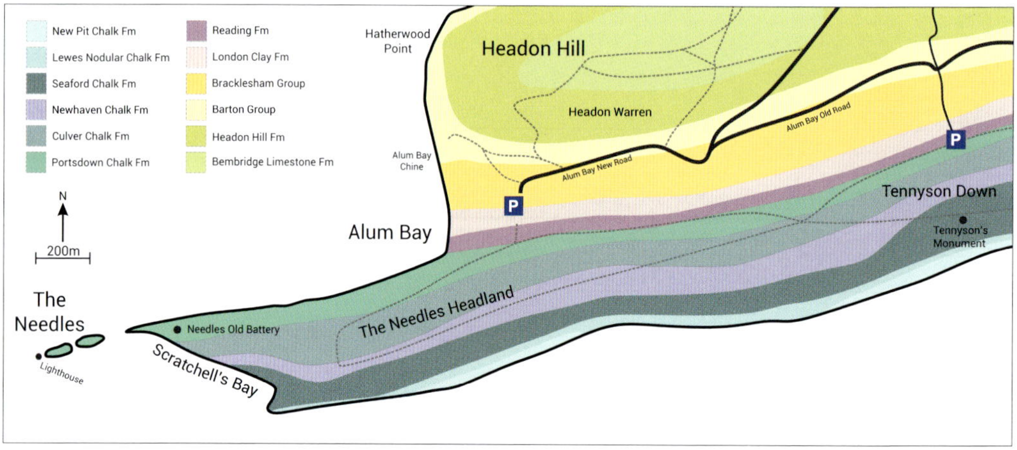

Figure 6.22

Compton Bay from Afton Downs.

dinosaurs, as well as the trackway and fossil forest (really a log jam), forming a reef at Hanover Point visible on a low tide. It is also the site of several major dinosaur discoveries, including *Vectiraptor* and a new giant spinosaur. Compton Bay has rocks spanning some 30 million years from the Wessex Formation floodplain through to the Chalk Sea deposits, representing a huge global sea level rise (*Fig. 6.22*).

Geology. Wessex Formation, Vectis Formation, Lower Greensand Group, Gault Clay Formation, Upper Greensand Formation, Chalk Formations and Pleistocene and Holocene river gravels.

Fossils. Bones and teeth of dinosaurs, pterosaurs, crocodiles, turtles, bony fishes and sharks, coprolites, molluscs, fir cones and flint fossils. Footprints of dinosaurs, mostly iguanodontids.

Access. Compton Bay and Hanover Point are easy to access, with a set of steps to access the beach from the large National Trust car park along the Military Road. The beach at Brook Bay is slightly more difficult due to recent land slipping. Parking is also available in the National Trust car park, near the junction of Military Road and Brook Village. There is a gravel track with a small set of steps and a walk across the small landslip to the beach provides access (*Fig. 6.23*). Both are situated on the no. 12 bus route.

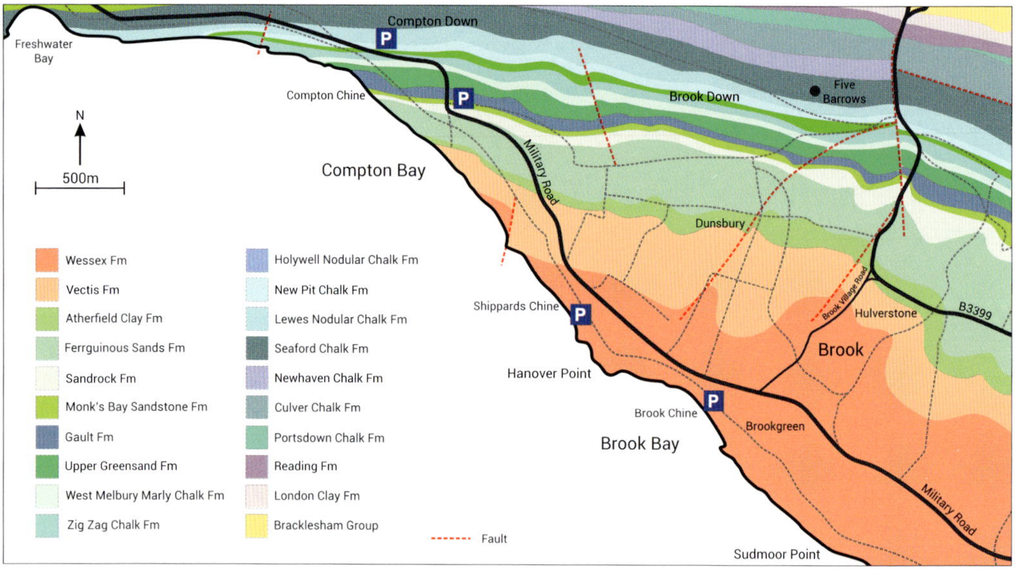

Figure 6.23

Map of Compton and Brook.

11. Chilton Chine

Chilton Chine is one of the most dinosaur-rich localities in all of Europe, with plenty of rolled bone to be found on the foreshore. It's also home to the new spinosaurs, *Riparovenator* and *Ceratosuchops*, as well as the well-known 'amber bed', to the east of the chine. A short walk west brings you to the impressive Sudmoor Point Sandstone, a mile-long and 10-metre-thick sandstone unit dominating the cliff. This sandstone is the remains of a huge river that once flowed across the island 128 million years ago (*Fig. 6.24*).

Figure 6.24

Chilton Chine.

Geology. Wessex Formation and Pleistocene river gravels.

Fossils. Bones and teeth of dinosaurs, pterosaurs, crocodiles, turtles, bony fishes and sharks, coprolites, molluscs, fir cones and flint fossils. Footprints of dinosaurs, mostly iguanodontids.

Access. Parking can be found in a small gravel car park along the Military Road, adjacent to the Isle of Wight Pearl.

A steep set of wooden steps to traverse, with the last portion being a series of steps cut into the cliffs, provides a short walk over the bottom of the landslip. The landslip is constantly moving, so access to the beach can be highly variable. The bus route no. 12 passes through the nearby village of Brighstone. It is approximately a 20 minute walk from the village to the beach, via New Road and the Military Road (*Fig. 6.26*). The Island Coaster bus passes by Chilton Chine twice a day between the 1st April and the 30th September (as of 2025).

12. Grange Chine

Grange Chine is probably the best place for flint fossils on the island, with the large shingle banks along the top of the shore providing a great number, especially echinoids. There is a high chance of finding pieces of dinosaur bones or fish teeth towards the bottom of the beach, close to the low tide mark. More complete

Figure 6.25
Grange Chine.

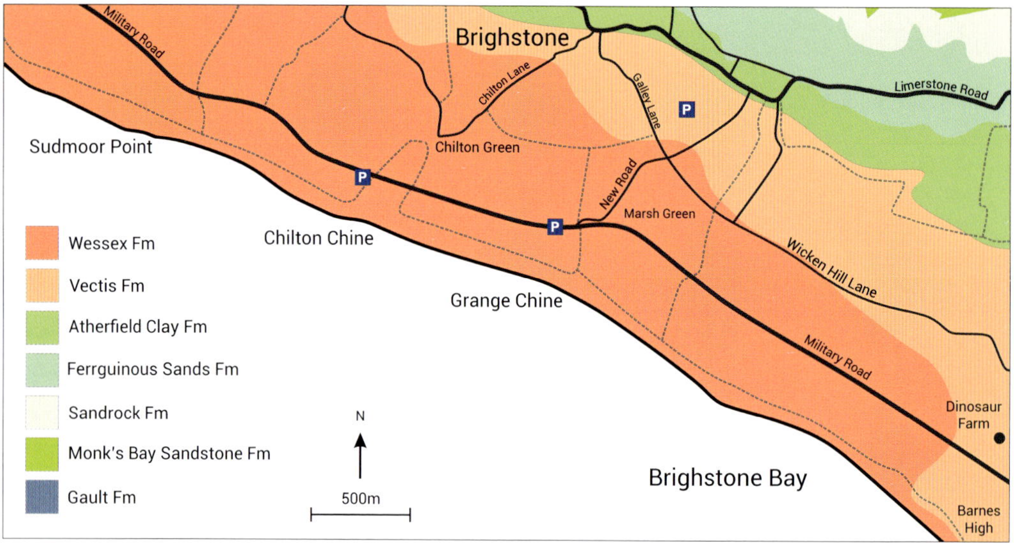

Figure 6.26

Map of Chilton and Grange.

dinosaur bones, mainly vertebrae, are also common along the stretch of beach between Grange and Chilton Chines (*Fig. 6.25*). Searching should be concentrated amongst accumulations of iron pyrite.

Geology. Wessex Formation and Pleistocene river gravels.

Fossils. Bones and teeth of dinosaurs, pterosaurs, crocodiles, turtles, bony fishes and sharks, coprolites, molluscs, fir cones and flint fossils.

Access. There is limited parking along the layby on the Military Road, and access to the beach is via a public footpath through the Grange Farm campsite. Additional parking can be found in the public car park in Brighstone village, with a short walk via a public footpath to Grange Chine (*Fig. 6.26*). There is a steep hill to climb up/ down and shingle banks to descend for access to the beach. Bus route no. 12 passes through the village of Brighstone, only a 10 – 15 minute walk to Grange Farm via public footpaths or along New Road. The Island Coaster bus passes by Grange Chine twice a day between the 1st April and the 30th September (as of 2025).

13. Atherfield to Blackgang Chine

Known locally as Chale Bay, this area is well-known for its large ammonites of the genus *Tropaeum*, and its so-called lobster beds, where complete crustaceans, called *Atherfieldastacus*, are commonly found. This part of the island's coastline is very fossiliferous, yielding all manner of marine fossils that once lived in a shallow sea, but also unfortunate dinosaurs washed out to sea and rare pterosaurs that dropped from the sky. Atherfield Point owes its existence to a highly fossiliferous sandstone rich in bivalves and known as the Perna Bed. It is incredibly hard and forms a prominent ledge extending over 1 km out to sea (*Fig. 6.27*).

The beach is often covered in deep shingle banks, which can be tiring to walk across. The cliffs here are incredibly unstable, with rock falls occurring almost daily. This site is not suitable for small children or those with limited mobility.

Geology. Vectis Formation, Lower Greensand Group.

Fossils. Rare dinosaurs, pterosaurs, turtles, bony fish, and shark teeth. More commonly echinoids, bivalves, ammonites, nautilus, brachiopods, arthropods, and corals in the Perna Bed.

Access. There is limited parking in the layby at Shepherds Chine. Access to the beach is either via the public footpath down through Shepherd's Chine, or along the coastal footpath to Atherfield Point and down the steps cut into the cliff by a local fisherman

Figure 6.27
Atherfield and Chale Bay.

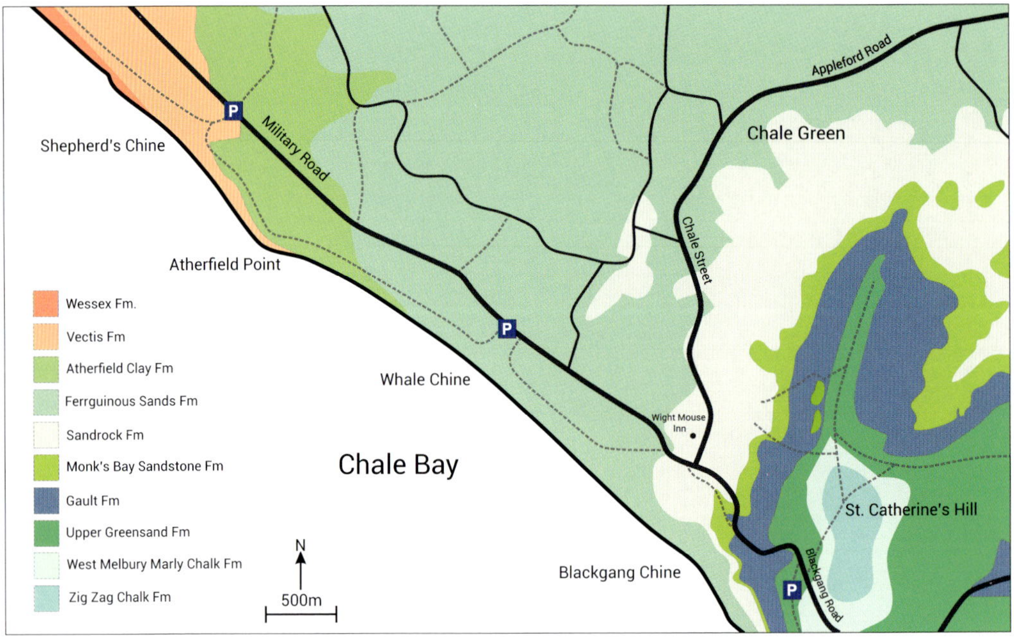

Figure 6.28

Map of Atherfield and Chale Bay.

and treasure hunter. A larger car park at Whale Chine provides access to the beach via the chine but is exceedingly difficult due to the stairs falling away years ago. New paths have been cut by local fishermen and beachcombers, but these can be difficult to traverse, especially after heavy rains. The closest bus route is the no.6 passing through Chale Village, a 10-15 minute walk to Whale Chine car park (*Fig. 6.28*). The Island Coaster bus passes by twice a day between the 1st April and the 30th September (as of 2025). The Wight Mouse Inn pub provides a welcome respite in bad weather, or a refreshing beverage after a day's collecting.

14. St. Catherine's Point/Rocken End

This area is an ancient landslip, which is still slowly moving south. The landslip has resulted in a jumble of blocks from 3 formations that have all tumbled down from the cliffs above (*Fig. 6.29*). Many of these blocks are highly fossiliferous, commonly yielding a high diversity of ammonites and other molluscs.

Fossils here are found amongst the beach shingle, but also within the large boulders on the beach. Hammers and chisels are enough to pop out fossils that can be seen on the surface. The beach

Figure 6.29
St. Catherine's Point/ Rocken End. Courtesy of Theo Vickers.

here is covered in large boulders and slippery rocks, so great care must be taken and is unsuitable for small children or those with limited mobility.

Geology. Gault Clay, Upper Greensand, Chalk.

Fossils. Marine reptiles, pterosaurs, shark teeth, bivalves, ammonites, nautilus, brachiopods and sponges.

Access. Parking can be found along St. Catherine's Road, with access via the track to St. Catherine's Lighthouse. A public footpath crosses the ancient landslip to Watershoot Bay (*Fig. 6.30*). There is no public transport to this locality. The No. 6 bus takes you into Niton, with a 45 – 60 minute walk to Watershoot Bay.

15. Shanklin

Shanklin is a nearly 3 mile stretch of beach and cliffs extending from Luccombe Chine to Sandown Pier and can be a highly fossiliferous locality, with pieces of fossil wood, shell nodules and ammonite steinkerns commonly found in abundance. This locality is now quite famous for yielding the predatory dinosaur *Vectaerovenator*.

All fossils are found loose on the foreshore the entire way along the beach. The cliffs at Shanklin are very steep, high and unstable, so a good distance should be kept from the base of the cliffs (*Fig. 6.31*).

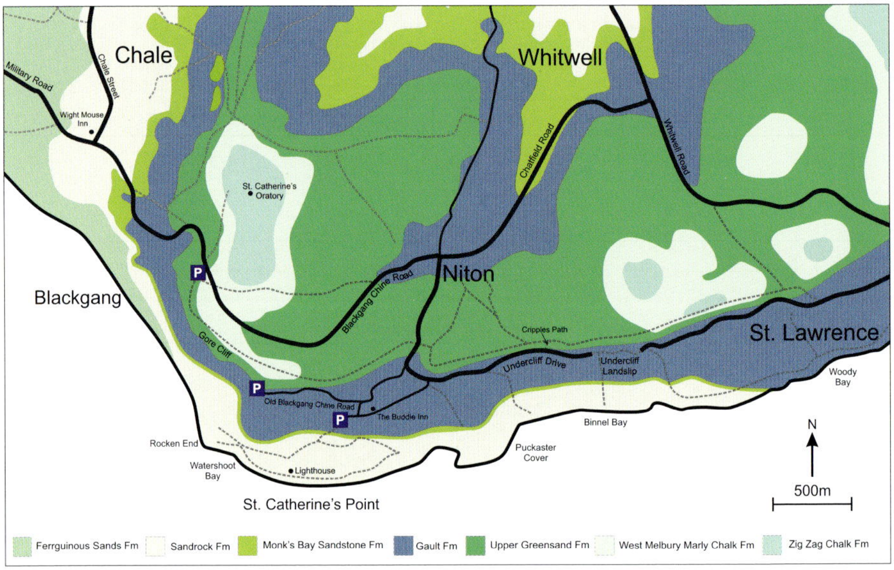

Figure 6.30

Map of Blackgang to St. Lawrence.

This is especially important at the southern end of the beach, where the cliffs are near vertical and highly unstable.

Geology. Lower Greensand.

Fossils. Rare dinosaurs and marine reptiles, ammonites, bivalves, brachiopods, sponges and fossils in flint.

Access. Shanklin is on the no. 2 and 3 bus routes and has ample parking. Access to the beach is via steps or slipways along the seafront, allowing for all ages and mobilities to get onto the beach safely and easily. There is also a public footpath along the seafront and the top of the cliff from Sandown (*Fig. 6.32*).

16. Yaverland

For the relatively small size of the exposure, Yaverland is amazingly fossiliferous, with around 30 million years' worth of geology exposed in just 1.5 kilometres of coastline. The rocks here yielded the pterosaurs *Caulkicephalus* and *Wightia* and the dinosaur *Yaverlandia*. The Vectis Formation slabs at Yaverland, located halfway between the beach access and white chalk cliff, are very fossiliferous being made of the bivalve *Filosina* or oysters. A close look at these slabs

Figure 6.31
Shanklin Bay.

will yield commonly *Hybodus* shark fin spines, bits of Jurassic ammonites and occasionally teeth of crocodiles and pterosaurs.

With easy access, and mostly a flat sand beach, Yaverland is a great locality for all ages. Although, for most of the summer months, the beach can sand up and finding fossils can be difficult (*Fig. 6.33*).

Geology. Wessex Formation, Vectis Formation, Lower Greensand Group, Gault Clay Formation, Upper Greensand Group, Chalk Formations and Pleistocene river gravels.

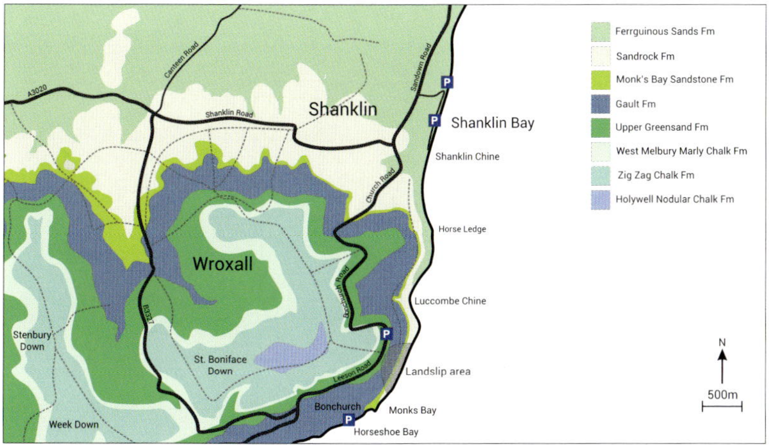

Figure 6.31
Map of Shanklin.

Figure 6.33

Yaverland showing red cliff and the chalk.

Fossils. Dinosaurs, pterosaurs, crocodiles, turtles, molluscs, sharks, coprolites, ammonites, bivalves, sponges and fossils in flint. Dinosaur footprints also occur but they are often covered with sand. The shaley parts of the Vectis Formation are rich in freshwater gastropods and microfossils, especially ostracods.

Access. There is ample parking along the esplanade and a carpark at the Yaverland end of the seafront (*Fig. 6.34*). Additional parking is also available at the Wildheart Trust. Access to the beach is either a small set of steps or down the slipway at the east end of Yaverland car park. The no. 2 and 3 buses will take you to Sandown, with a short walk to Yaverland. The no. 8 bus stops at Yaverland. There is an excellent café next to the car park and public toilets in the car park too.

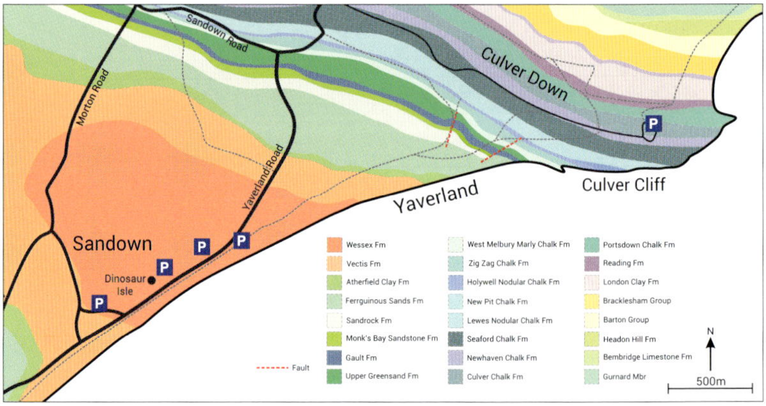

Figure 6.34

Map of Yaverland.

17. Whitecliff Bay

There are very few localities in northwest Europe that rival the Paleogene geology of Whitecliff Bay, with exposures of almost the entire sequence of Paleogene strata in the Hampshire Basin. Spanning the late Paleocene through the Eocene to the early Oligocene Bouldnor Formation, approximately 20 million years of Paleocene and Eocene rocks can be explored here.

Situated to the south of Bembridge along the northeast coast of the island, Whitecliff Bay is always open and usually sandy. The prominent chalk headland of Culver Downs forms the southern end of the bay with cliffs of overlying Paleogene strata extending northwards toward Bembridge. In the southern part of the bay, the strata are vertical and change character rapidly as you walk northwards from the chalk cliff. Standing on the water line on the lowest tides, or from a boat offshore, the cliffs have a wonderful striped appearance (*Fig. 6.35*). There is a café at beach level below the caravan park.

Geology. Chalk Group, Reading Formation, London Clay, Bracklesham Group, Barton Group, Headon Hill Formation, Bembridge Limestone Formation.

Fossils. Eocene mammals, turtles, alligators, fishes, bivalves, gastropods, arthropods, serpulid worms, giant foraminifera and Cretaceous flint fossils. The chalk headland is rich in fossil echinoids.

Figure 6.35
Whitecliff Bay.

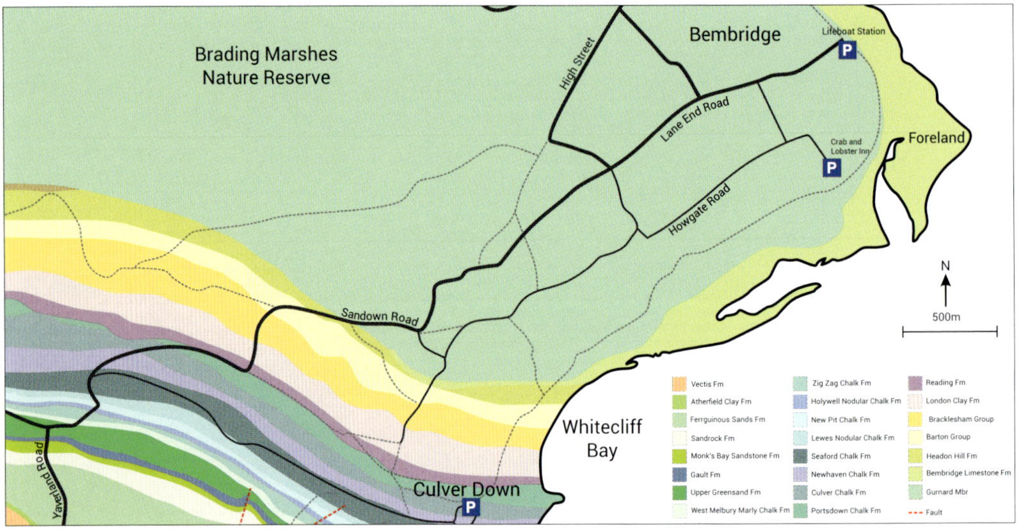

Figure 6.30

Map of Whitecliff Bay to Bembridge.

Access. Car parking can be found in Bembridge at the end of Lane End Road and the end of Forelands Field Road. Access to the beach from these points is down a slipway and set of steps, respectively. Accessing via the Whitecliff Bay holiday camp has little parking available, with steep ramps and steps from the caravan park (*Fig. 6.36*). The no.8 bus, along with the Downs and Island Breezer buses serve the Bembridge area.

7. Fossil Identification

Here are the most common fossils (and some beautiful, rarer extras) you can find on the island's beaches, starting with the oldest, with lots of information about the palaeobiology of the animals you may stumble across.

This is by no means a complete guide to all the fossils the island yields, but a starting point for the casual collector walking our shores. If you find something that's not within this guide, it's always good to show your finds to a local museum, university geology department or island fossil experts. Rare and new species are discovered quite frequently, so it's always worth getting your finds checked out. The main starting point for identification is where you found the fossil, this narrows the search down to the rocks exposed at that locality. This can then be followed using the colour, texture, and shape of the fossil to identify it further.

7.1. Fauna of the Wessex Formation floodplains

The Wessex Formation yields a remarkably high abundance of fossils in a relatively small amount of exposure, with the most common finds being rolled bone and fossil wood on the foreshores. However, with enough time, dedication and luck, many amazing finds can be made, including on rare occasions, almost complete skeletons!

Molluscs

The Wessex Formation rarely yields fossil molluscs (snails and clams). It is thought that the sediments they were buried in were slightly acidic, such that their shells have long since dissolved away. Although we know they must have been a crucial part of the ecosystem, fish and crocodiles have been found with teeth that are specialised for crushing shells. However, one plant debris bed between Brook and Chilton does yield large unionid bivalves called *Margaratifera valdensis* (*Fig. 7.1*). These are very distinctive large freshwater mussels often found amongst the shingle. A smaller unionid occurs in one of the plant debris beds at Yaverland.

Interestingly, at both localities, the bivalves are found with both shells tightly closed, and it is thought they were dislodged from the riverbed during flooding and redeposited in mud further downstream. Unfortunately, they were buried too deeply and suffocated before being able to burrow back to the surface.

Possible snail eggs

Some fragments of bone have surfaces covered in small black spheres or just the impressions of where the spheres once were. These have been interpreted as gastropod (snail) eggs. However, the exact origin of these is still debated. Similar structures are also found inside some of the unionid bivalve fossils (*Fig. 7.1*).

Figure 7.1

Molluscs of the Wessex Formation.
A, *Margaratifera valdensis*.
B, 'snail eggs' under 200x magnification.
C, *Margaratifera* bivalve with 'snail eggs'.
A, collection of Huxley Hunt.
Scale equals 10 mm.

Fish

Fish remains are common amongst the finer shingle at the base of the beach, these include diamond-shaped scales and round button teeth from the durophagous (shell-crushing) fish *Scheenstia*, formerly called *Lepidotes* (Fig. 7.2).

Figure 7.2

Fish of the Wessex Formation. A, pycnodont palette. B, E, D, F, *Scheenstia* tooth palettes. C, G, isolated scales. A, F, Dinosaur Expeditions C.I.C collections. Scale equals 10 mm.

Dinosaurs

Teeth

Teeth can also be found amongst the shingle on the beach. Mostly black or deep brown with a shiny surface. The base of the tooth will

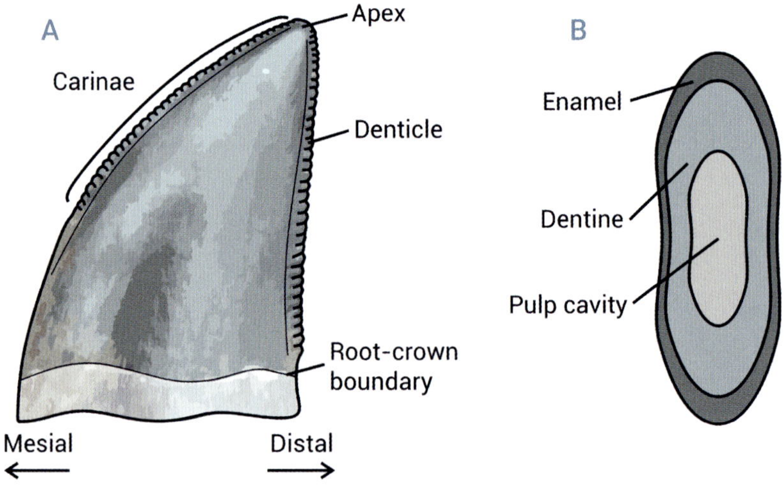

Figure 7.3

Diagram showing the structure of a tooth, labelled with standard scientific terminology.

show 3 concentric rings, which are the enamel, dentine, and pulp cavity, which is often a central cavity in the tooth (*Fig. 7.3*). It is usually possible to identify the original owner of the tooth if it's not too beach rolled (*Fig. 7.4*).

Teeth of the carnivorous theropod dinosaurs, such as *Neovenator*, are recurved and blade-like with a narrow oval cross-section. The teeth towards the front tend to be more robust, and a little fatter than those from the back of the jaws.

Teeth of herbivorous dinosaurs like *Iguanodon* are highly distinctive being leaf-shaped with 2 distinct ridges and serrated margins. There is usually a wear facet at the top of the tooth, caused by the animal grinding the tooth down by chewing on tough vegetation.

Teeth of spinosaurids are often confused with crocodile teeth, as both have similar shaped teeth, ideal for eating fish. Until recently, with the discovery of *Baryonyx*, most of the spinosaurid teeth from the island were identified as crocodiles. Both are very conical and

Figure 7.4

Wessex Formation dinosaur teeth. A, spinosaurid isolated crown. B, ?*Vectiraptor* isolated crown. C, ?*Eotyrannus* sp. isolated crown. D, ?*Neovenator* isolated crown. E, spinosaurid isolated crown. F, iguanodontid maxillary crown. G, iguanodontid dentary crown. H, digital drawing of a *Polacanthus* tooth. I, beach rolled shed iguanodontid crown. J, iguanodontid dentary crown. K, peg-like sauropod crown. L, *Oplosaurus* sauropod tooth. A, collection of Kai Bailey. E, L, collection of Paul Batty. B, D, G, F, J, collection of Mick Green. I, collection of Sam Caethoven, K, collection of Rubens Braun. A, C, D, E, F, G, I, J, K, L, scale equals 10 mm. B, H, scale equals 5 mm.

can be recurved. However, crocodile teeth have the carinae (ridge which can have serrations on), on the sides opposite to the curve. Whereas spinosaurid teeth have the carinae on the sides with the curve.

Figure 7.5

Dinosaur bone.
A, black bone. B, brown bone from the red beds. C, blue and orange bone from the mottled marls. D, white bone. E, red bone. Scale equals 10 mm.

Bone

Dinosaur bone often has a characteristic honeycomb texture seen on broken or worn surfaces. Most of the dinosaur bone found is black or very deep brown and is often quite heavy for its size. The dark colour is likely due to a high iron content within the bones, with pyrite infilling micro voids within the bone's microstructures, adding extra weight to the bone.

It's also possible to find bone in a variety of different colours (*Fig. 7.5*). These often brightly varicoloured bones come from the mottled clays and red beds and take on the colour of the clay. In the red beds, the bones may be bright white or creamy white like modern bones. Likely these bones are completely oxidised and have little or no organic material left within them. Such bones from the mottled clays and red beds tend to be extremely fragile and do not survive long on the beach.

With practice, it is possible to identify which bone of the skeleton you have found based on the shape. Here are some examples of the most common bones that occur on the island.

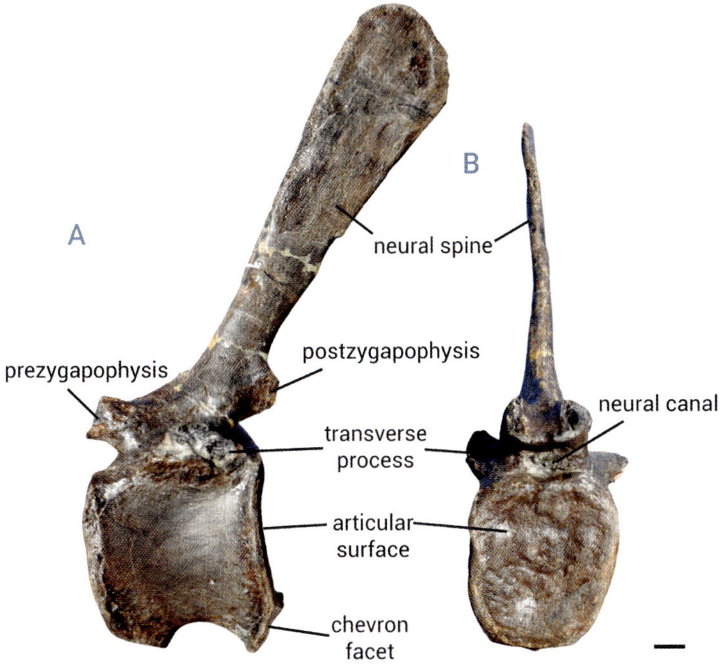

Figure 7.6

Ornithopod caudal vertebrae labelled in A, lateral and B, anterior views. Dinosaur Expeditions collection.

Vertebrae

Vertebrae have two distinct parts, the centrum, and the neural arch (*Fig. 7.6*). It is common to find rolled centrums on the beach or fragments of the neural arch.

Vertebrae are categorised by their position along the vertebral column (*Fig. 7.7*). Neck vertebrae are called cervicals, while those along the back are called dorsal vertebrae and those from the lower back are called lumbars. Where the vertebral column passes through the pelvic region the vertebrae may become fused (and parts of

Figure 7.7

Diagram of an *Iguanodon* skeleton.

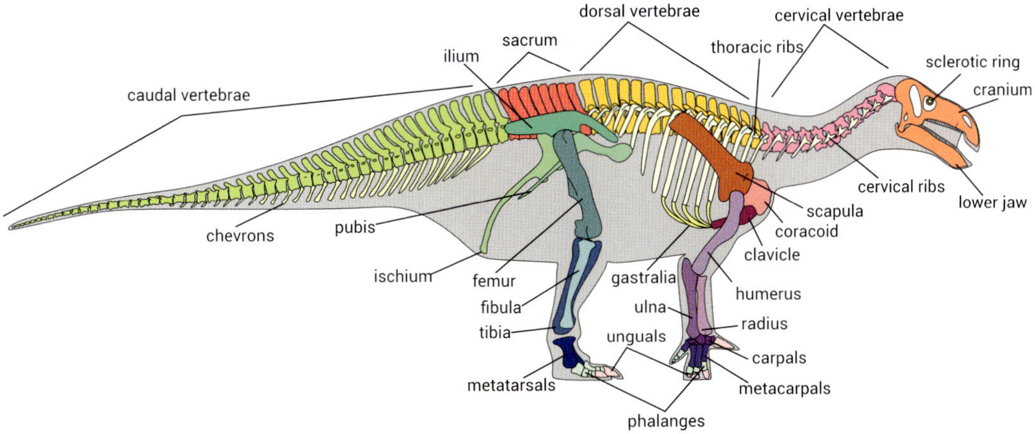

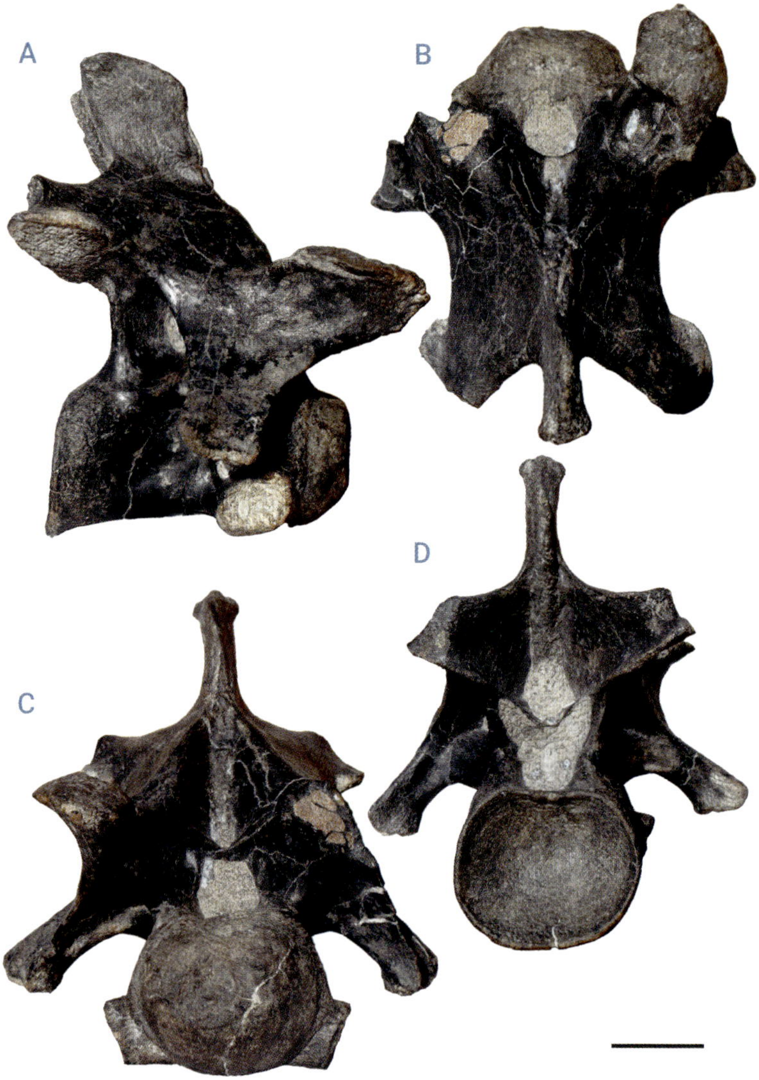

Figure 7.8

Spinosaurid cervical vertebrae in A, posterior, B, anterior, C, lateral and D, dorsal view. Collection of Kai Bailey. Scale equals 50 mm.

the pelvic girdle), known as a synsacrum or sacral vertebrae. The number of centra fused can vary between dinosaurs, but three appears to be a minimum number. The vertebrae of the tail are called caudals, and they may have chevrons attached. These are spines that sit underneath the vertebral column and are attached at the junction between two caudal vertebrae.

The shape of the vertebra differs depending on where the bones come from along the backbone, and from which animal they come from too. Cervical vertebrae often have complex shapes and are

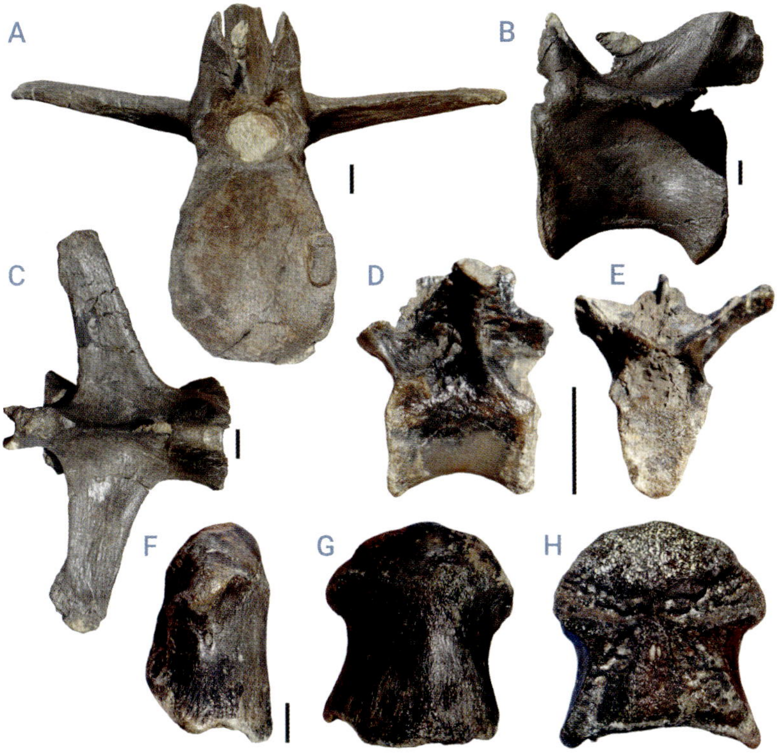

Figure 7.9

Examples of theropod vertebrae. Spinosaur caudal vertebrae in A, anterior, B, lateral and C, dorsal view. ?*Calamosaurus* dorsal vertebrae in D, lateral and E, anterior view. ?Spinosaur cervical centrum in F, lateral, G, ventral and H, dorsal views. A,B,C, collection of Kai Bailey. D-H, collection of Dinosaur Expeditions C.I.C. Scale equals 10 mm.

slightly elongate (*Fig. 7.8*). Dorsal vertebrae are more robust and have a chunky hourglass shape.

Caudal (tail) vertebrae differ along the tail. At the top of the tail, the centra can resemble the dorsal vertebrae. Going down the tail, a notch develops on the front and back of the base of the centrum, for a chevron to articulate with.

This notch is where the chevron would have articulated. Towards the end of the tail the bones become very round in cross-section, with increasingly smaller neural arches and end up looking like cotton reels.

Theropods have very narrow, slender centra. Often very pinched in the middle forming an hourglass shape. Although there is variation along the vertebral series, the centra remains very slender and pinched (*Fig. 7.9*).

Ornithopods (herbivorous dinosaurs like *Iguanodon* and *Hypsilophodon*) have chunkier centra and are the most commonly occurring vertebrae in the Wessex Formation. With a circular to oval

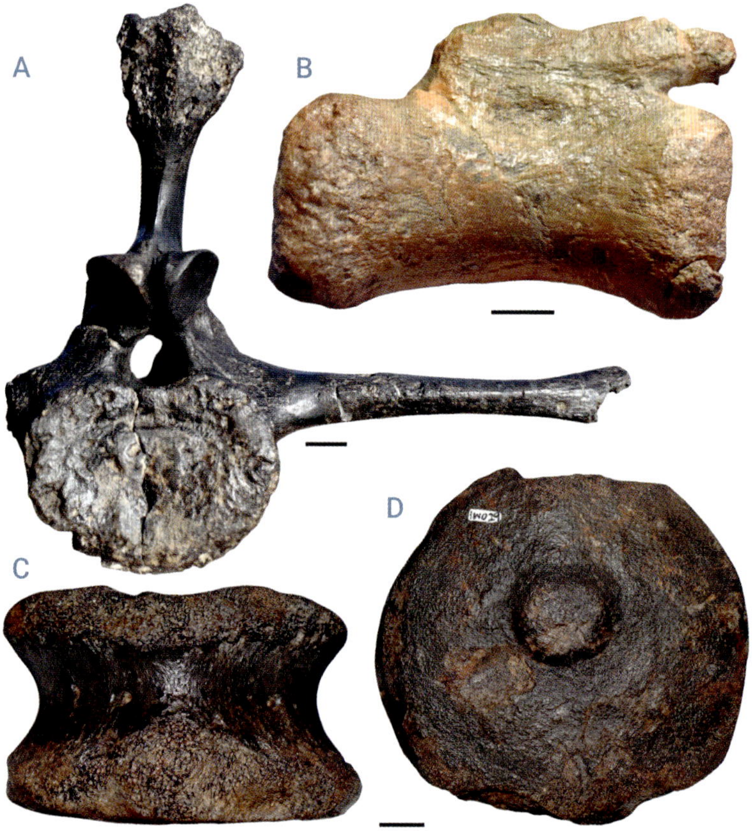

Figure 7.10

A, *Polacanthus* vertebra. B, sauropod caudal vertebra. C, D, sauropod dorsal vertebra in ventral and anterior view. A, collection of Dinosaur Expeditions C.I.C. Scale equals 10 mm.

cross-section and only a slight pinching in the middle. However, the vertebrae vary considerably along the vertebral column. Mid caudal (tail) can have a hexagonal cross-section, and the centra towards the end of the tail look like cotton reels.

Sauropods (long-necked dinosaurs like *Brachiosaurus*) have almost hollow vertebrae, especially in the neck, allowing them to grow to gigantic sizes. However, this makes their bones fragile. Usually the cervical (neck) vertebrae are subject to crushing but are very elongate and complex. The dorsal vertebrae are normally exceptionally large, with a circular cross section and are noticeably short in length (*Fig. 7.10*). Caudal vertebrae are elongated with a more rectangular cross section.

Ankylosaurid (armoured dinosaurs like *Polacanthus*) vertebrae may resemble those of ornithopods, however, a key difference is a deep, wide neural canal and a more rounded heart-shaped

Figure 7.11

A, dinosaur rib with reconstructed articular end, B, cross section of a dinosaur limb bone. C, rib cross section showing growth lines within the bone structure. D - G, examples of chevrons. Scale equals 10 mm.

cross section, usually with a centrum slightly wider than it is tall (*Fig. 7.10*).

Limb Bones

Limb bones are usually circular in cross-section, often with a hollow cavity (medullary cavity) inside, that is usually infilled with calcite and pyrite. The morphology of the articulatory ends of limb bones vary depending on the bone and can be diagnostic (*Fig. 7.11*).

Ribs

Rib bones can be quite diagnostic for general identifications. Features to check include the number of articulations, the cross-sectional shape and the presence or absence of flanges. The degree of curvature may also be helpful. In dinosaurs, ribs are found on the cervical and thoracic vertebrae. They also occur on the sacral vertebrae where they are usually fused to the centrum and the pelvis. Very thin ribs that are pointed at both ends may be found to form a girdle between the thorax and pelvis. These ribs are called gastralia.

Sections of rib have a characteristic triangular cross section on the broken surfaces. The cross section becomes flatter the further down the rib. The top of a rib, which articulates with the vertebrae is robust, and is V shaped (*Fig. 7.11*).

Chevrons

Chevrons are V-shaped bones from the tail skeleton, articulating with the bottom of the caudal centra and the junction between

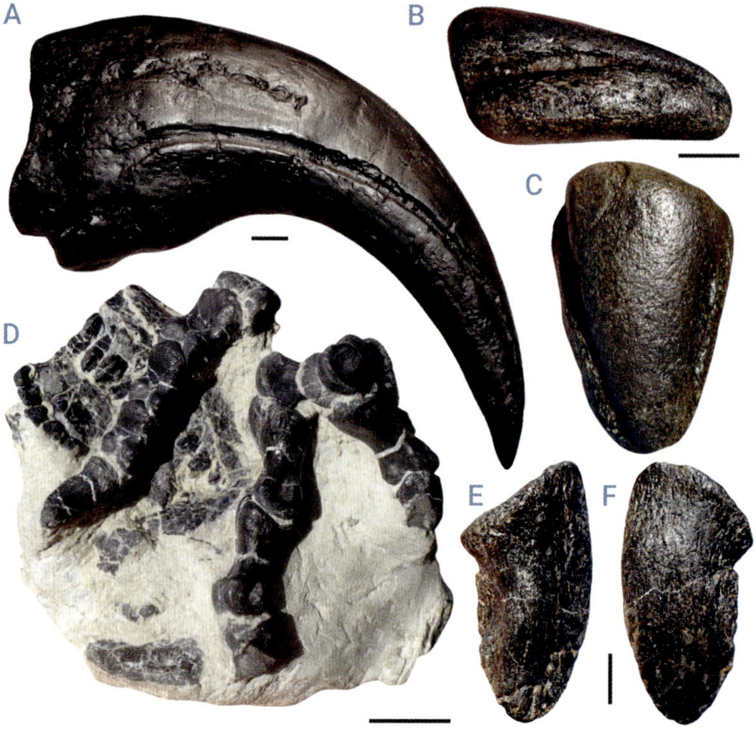

Figure 7.12

Dinosaur claws and phalanges. A, cast of a *Baryonyx* hand claw. B, C, ornithopod foot claw. D, *Hypsilophodon* foot. E, F, Sauropod foot claw. D, collection of Kai Bailey. E, F, collection of Dinosaur Expeditions C.I.C. Scale equals 10 mm.

the vertebrae (*Fig. 7.11*). They are sometimes called haemal arches (as opposed to neural arches) because large blood vessels pass through them.

Phalanges

The phalanges are the finger and toe bones. These can be mistaken for vertebrae by the untrained eye. However, there are a few crucial differences to look out for; phalanges are usually flatter on their base than on their upper surface. They are generally wider at their proximal end than their distal end, and there is a conspicuous 'roller joint' at the distal end with a circular depression laterally (*Fig. 7.12*).

Unguals

Unguals or claw bones can range from being extremely curved, like the classic '*Velociraptor*' toe claw of Jurassic Park fame, to flat, triangular claws depending on which animal it comes from, and whether it's a foot or hand claw. Theropod and sauropod claws tend to be more curved and pointed, whereas *Iguanodon* claws are wider,

Figure 7.13

1881 drawing by G. M. Woodward of the NHMUK R175 *Polcanthus* sacral shield found by Rev. William Fox.

flat, and blunt, some curving sideways rather than forwards. Unlike typical dinosaur claws, *Iguanodon* thumb spikes are large, sharply pointed, and conical unguals. Most reptile claws have a distinct groove on either side where the claw sheath is embedded (*Fig. 7.12*).

Armour

On rare occasions pieces of dinosaur armour appear on the beach. These are mostly attributed to a thyreophoran dinosaur called

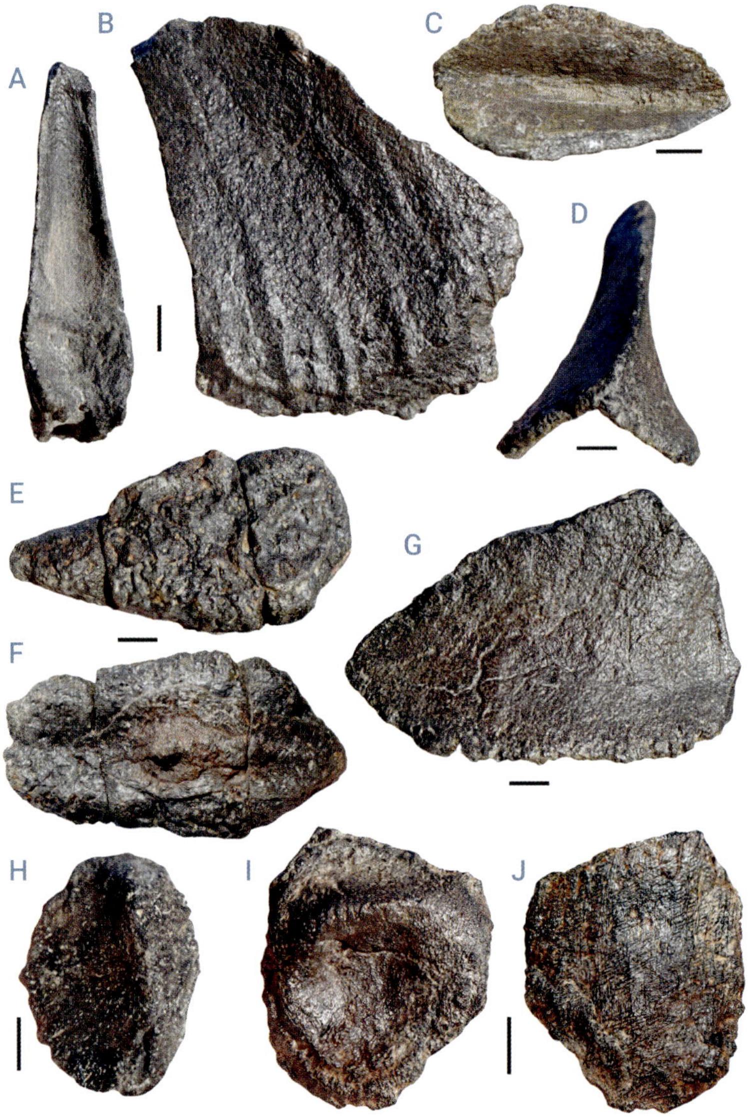

Figure 7.14

Examples of *Polacanthus* armour. Caudal plate in A, posterior, B, lateral and C, ventral view. Dermal scute in E, lateral and F, ventral view. Caudal plate in D, anterior and G, lateral view. Dermal ossicles in H, I, dorsal and J, ventral view. A-J, collection of Dinosaur Expeditions C.I.C. Scale equals 10 mm.

Polacanthus or the newly named, *Vectipelta*, a relative of the famous late Cretaceous *Ankylosaurus*. *Polacanthus / Vectipelta* had a bony shield over its hip region (*Fig. 7.13*) and was adorned with spines and plates along its back, legs and tail. These pieces are very characteristic, having a woven fabric bone texture on the surface attached to the skin and can have grooves over the surface, which are for the accommodation of the blood vessels between the bone and overlying horny scute (*Fig. 7.14*).

Ossified tendons

Ossification is the natural process of bone formation and the hardening of tissues into a bony substance.

Ossified tendons are found in ornithischian dinosaurs, like *Iguanodon*, and have been classed as a defining character. However, they've also been found in theropods and pterosaurs, and even crop up in extant turkeys. They are typically found forming a tendon

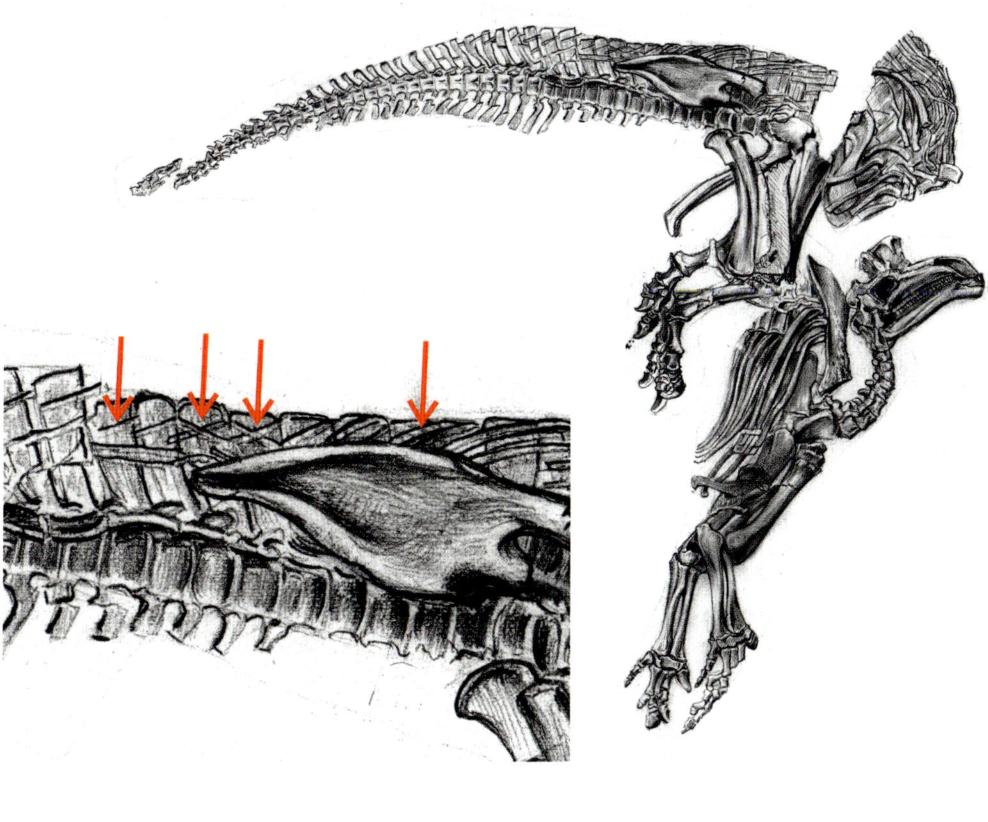

Figure 7.15

Drawing of one of the Bernissart *Iguanodon* specimens by G. Lavalette, 1882, showing the ossified tendon lattice across the vertebral column.

lattice supporting the sacral vertebrae but have also been found on the dorsal and caudal vertebrae too, where they form a criss-cross lattice along the sides of the vertebrae. These ossified tendons significantly stiffen the vertebral column over the hips and reduce the amount of movement in the upper part of the tail (*Fig. 7.15*).

Ossified tendons can be found as rolled pieces on the foreshore and are quite distinctive. Look out for rod-like pieces of bone, with a very dense honeycomb texture. Usually quite fibrous with needle-like dots on the broken ends.

Aging Dinosaurs

It is possible to estimate the age of a dinosaur by assessing the amount of fusion between the bones, such as in the skull (*Fig. 7.16*) or between the centrum and the neural arch of the vertebrae. In

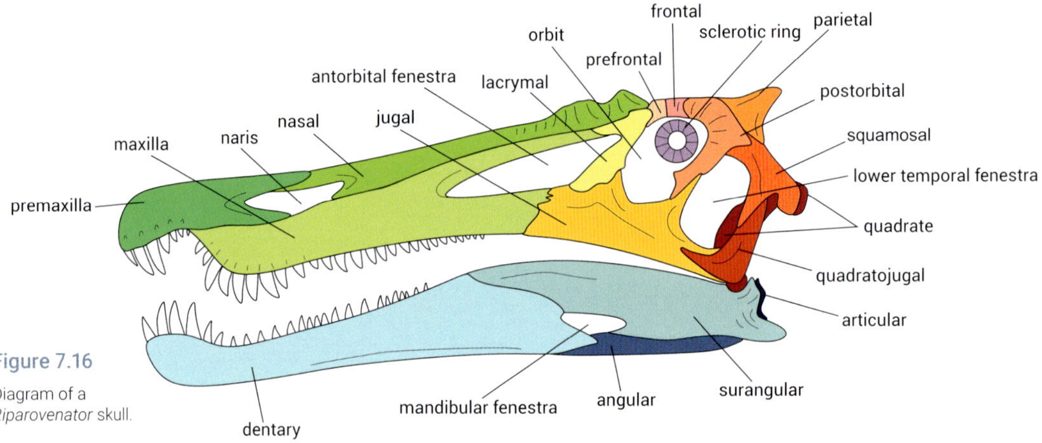

Figure 7.16

Diagram of a *Riparovenator* skull.

young animals, there is no fusion, and these remain as two separate bones. The top of the centrum and the bottom of the neural arch have a unique surface texture of ridges and grooves (*Fig. 7.17*). If this is present, it's evidence of a young animal. As the animal matures, these bones slowly fuse. In a young adult, the two bones may be fused, but still display a clear line showing the boundary. In old individuals, the fusion is usually complete and there may be little in terms of a boundary line between the two bones.

Scientists can also count the lines of arrested growth (LAG lines) in bone to approximate a dinosaur's age. LAG lines are the remains of old bone walls within the bones, often looking like a series of prominent concentric growth lines (*Fig. 7.18*). As the dinosaur grows, it lays down a new outer bone layer and slowly reabsorbs the older inner layers. This process leaves remains of the bone wall inside the bone, which are thought to be annual or seasonal growth bands. It is quite reminiscent of the rings in tree stumps. These LAGs are best seen in thin sections of the bone (a slice on a glass slide, polished to around 0.003 mm thick) but can be seen on the outer margins of broken bones. Due to the amount of remodelling the bone undergoes as the animal grows, usually, all the juvenile bone has been completely reabsorbed, so it can only provide a minimum age of the animal, not the true age.

Figure 7.17

Left: Vertebrae showing various levels of fusion. A, unfused centrum with ridged and grooved texture. B, fused centrum and neural arch showing the fusion boundary. C, complete fusion of neural arch and centrum, with boundary no longer visible. Scale equals 10 mm.

Gastroliths

Gastroliths, also called stomach stones or gizzard stones, are rocks swallowed by the animal and held within the gastrointestinal tract to aid in digestion, or in the case of some marine reptiles, to aid buoyancy. Gastroliths found in the Wessex Formation have been suggested to come from sauropod dinosaurs. These dinosaurs cannot chew their food, having only peg-like teeth at the front of the jaws with no molar teeth to grind down tough plant material. It's likely sauropods simply swallowed everything whole. Therefore, they needed to swallow stones to help them grind up and digest the tougher plant material while it was in the gut.

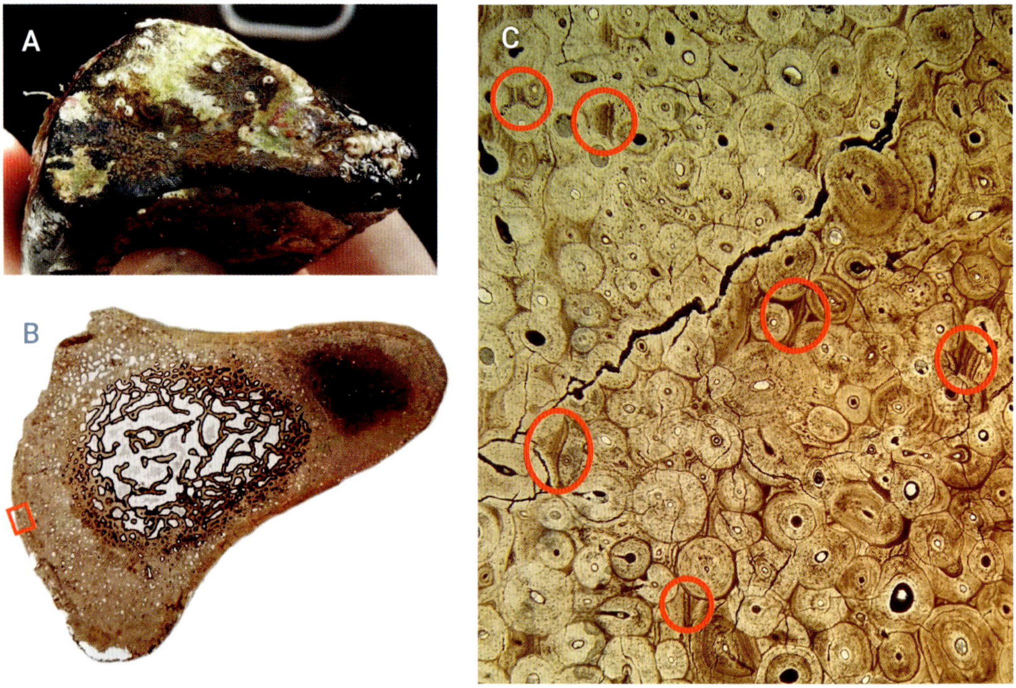

Figure 7.18

Thin section of bone showing LAGs (lines of arrested growth).
A, photo of a rib section before sectioning. B, scan of thin section showing the area in C. C, view of bone under the microscope, with red circles highlighting LAGs.

Gastroliths are large, smooth pebbles with a high polish. The polish is created while the gastrolith was inside the dinosaur's gizzard, with the polishing agent being the microscopic silica grains (called phytoliths) in the plant material the dinosaur ate.

On the Isle of Wight, gastroliths are found sticking out of plant debris beds, often being able to pull them out with little effort.

Figure 7.19

Examples of gastroliths found within the Wessex Formation.
A, haematite. B, quartzite. C, altered vein quartz. D, tourmalised sandstone. Scale equals 10 mm.

Once the stones are on the beach and have been rolled by the sea, it becomes impossible to identify them as gastroliths as they lose their characteristic polish (*Fig. 7.19*).

Crocodiles

The Wessex Formation is home to at least 6 different types of crocodile, including the specialist mollusc eater, *Kompiodontosuchus*. The name translates to 'button tooth crocodile', in reference to its squat, rounded teeth used to crush hard-shelled prey. The

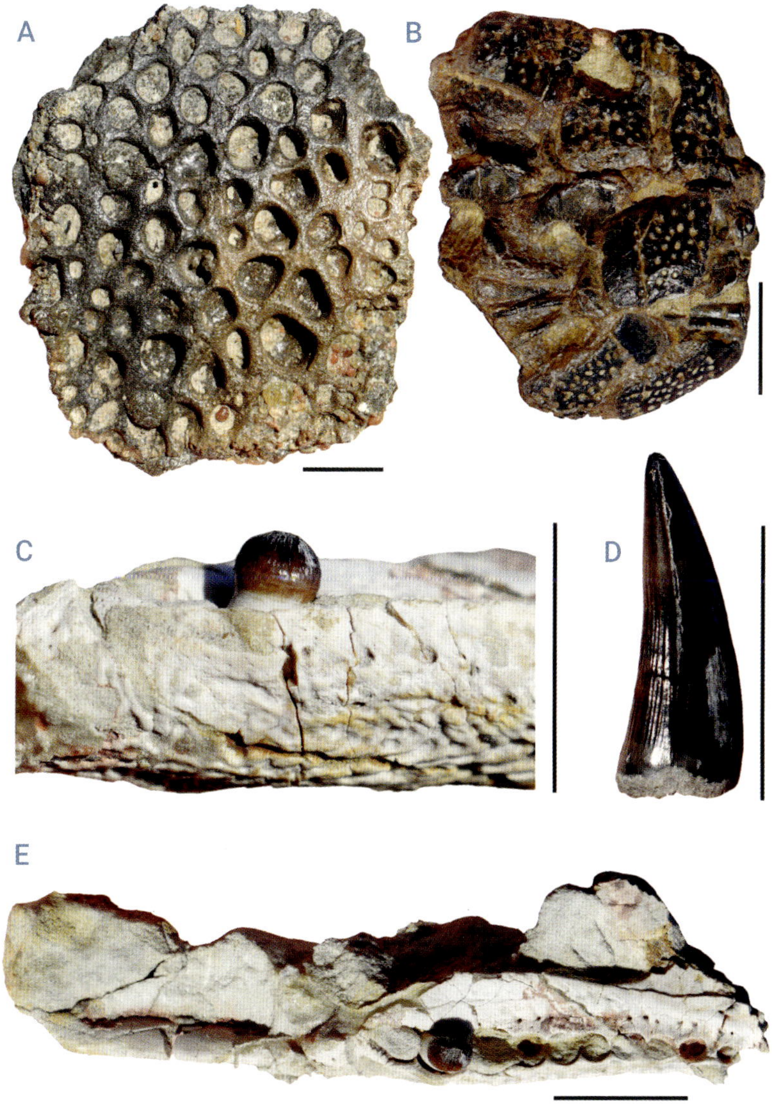

Figure 7.20

Wessex Formation crocodyliforms. A, large scute. B, partially articulated crocodyliform. C, *Kompiodontosuchus* tooth in jaw. D, goniopholid tooth crown. E, right dentary of *Kompiodontosuchus* in dorsal view. A-E Dinosaur Expeditions C.I.C collection. Scale equals 10 mm.

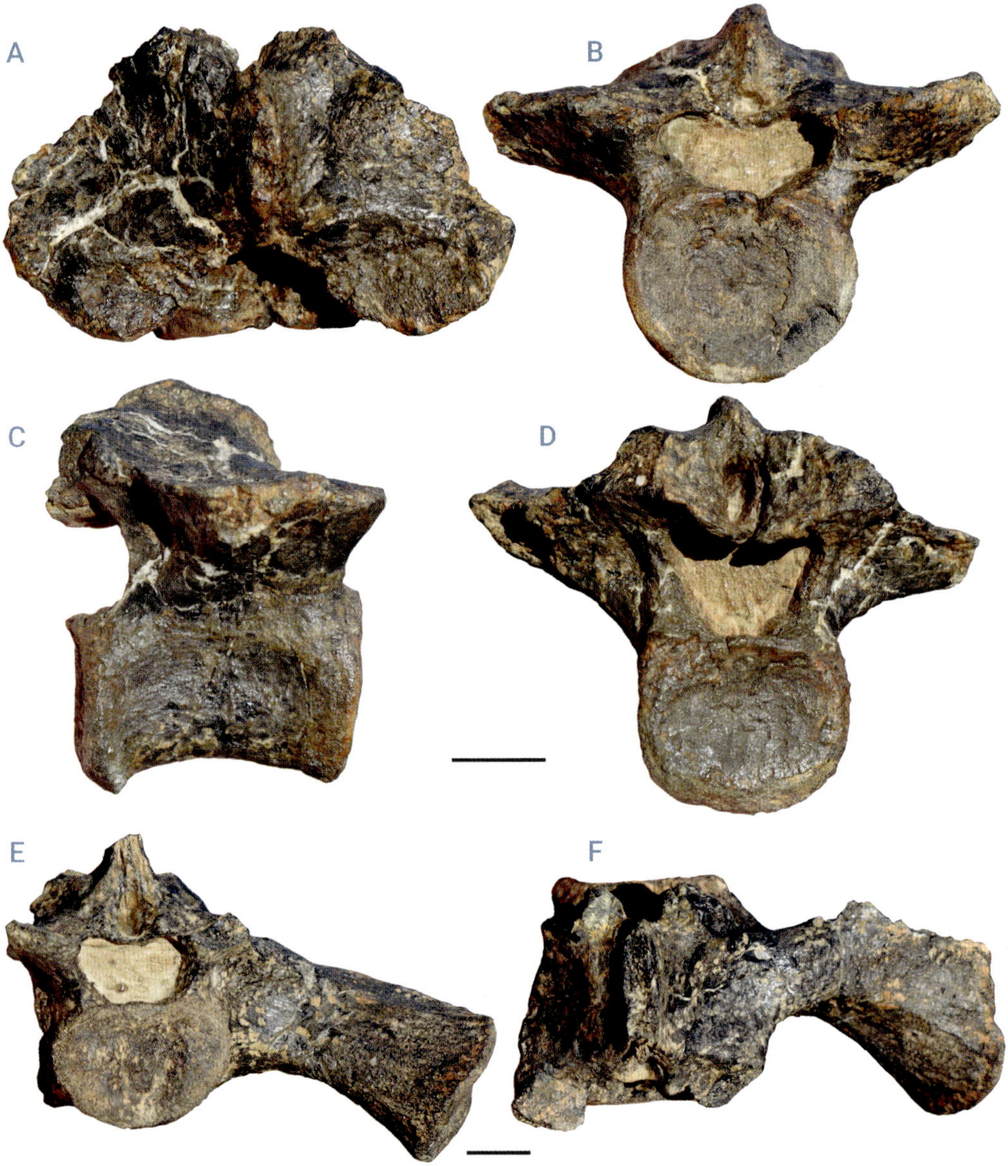

largest crocodile present in the formation is *Anteophthalmosuchus* (formerly known as *Goniopholis*), coming in at over 3.5 metres long (*Fig. 7.20, 7.21*).

Figure 7.21

Goniopholid dorsal vertebra in A, dorsal, B, anterior, C, lateral and D, posterior views. Goniopholid sacral vertebra in E, anterior and F, dorsal view. Collection of Dinosaur Expeditions C.I.C. Scale equals 10 mm.

Crocodile osteoderms and skull bones are easy to identify, by having a unique pitted pattern on the surface. Osteoderms are the bony plates that lay under the crocodiles' skin during life, acting as a form of support for the armour of often sharp, horny scutes. The pattern of these osteoderms can be diagnostic between species.

Crocodile teeth are also a relatively common find and easy to identify. The tooth crowns are conical, sometimes bearing a raised ridge (carinae) along the sides, rather than front and back as seen in theropod dinosaurs (see the section on dinosaur teeth). The carinae may have very fine serrations on the carinae in well-preserved specimens.

Turtle

The Wessex Formation yields a variety of different turtles, including the terrestrial turtle, *Helochelydra nopcsai* and the freshwater turtles *Brodiechelys brodiei*, *Eodortoka* cf. *morellana* and *Pleurosternon* sp.

Turtle shell fragments are easily identified as thin pieces of bone with a dense bone structure, with some pieces having a unique pattern over the convex surface, which can be dots or

Figure 7.22

A, example of *Helochelydra* shell showing B, unique sphere ornamentation. C, photograph and D, rendered micro-CT image of 'Burby' in dorsal view. Scale equals 10 mm.

ridges. The inner, concave surface is almost always smooth, and even shiny. And on rare occasions, skulls or even whole turtles turn up. One such example is 'Burby', a small turtle with all the bones preserved inside the shell found by local fossil collector, Mr Steve Burbridge (*Fig. 7.22*).

Botany

The Wessex Formation yields a high diversity of plant material, ranging from massive logs, branches, twigs and leaves to beautiful 3D pine and cycad cones. The wet conditions of the floodplain, intense forest fires, with rapid burial of remains have led to the plant material being well preserved. Without the rapid burial into anoxic sediments, such organic material simply decays away. Although it's mostly overlooked by many collectors, the preservation of the botanical material is exceptional within the Wessex Formation.

Figure 7.23

Section of intensely burnt carbonised wood from a plant debris bed fall at Brook.

Plant Debris Beds

Fossil wood is extremely common in plant debris beds. The wood ranges from intensely burnt wood from intense forest fires, including a powdery charcoal called pyrofusinite, and pieces which have simply been washed onto the floodplain and buried without burning (*Fig. 7.23*). Most of the fossil wood is black, often with pyrite (fool's gold) within the grain of the wood or as an outer coating. There is often a greenish tinge to the pyrite if it's been on the beach for a length of time (*Fig. 7.24*). On rare occasions, leaves can be found on split sections of plant debris bed.

Figure 7.24

A, B, examples of Wessex Formation pinecones. C, polished piece of Wessex Formation amber.
D, rolled pieces of wood with pyrite. Scale equals 10 mm.

Pinecones

Conifers and cycad cones are common, either pyritized, compressed or sometimes found within sandstone. These often have a characteristic diamond pattern over the surface which are the scales of the cone. These can vary depending on the preservation, degree of abrasion and the type of cone (*Fig. 7.24*).

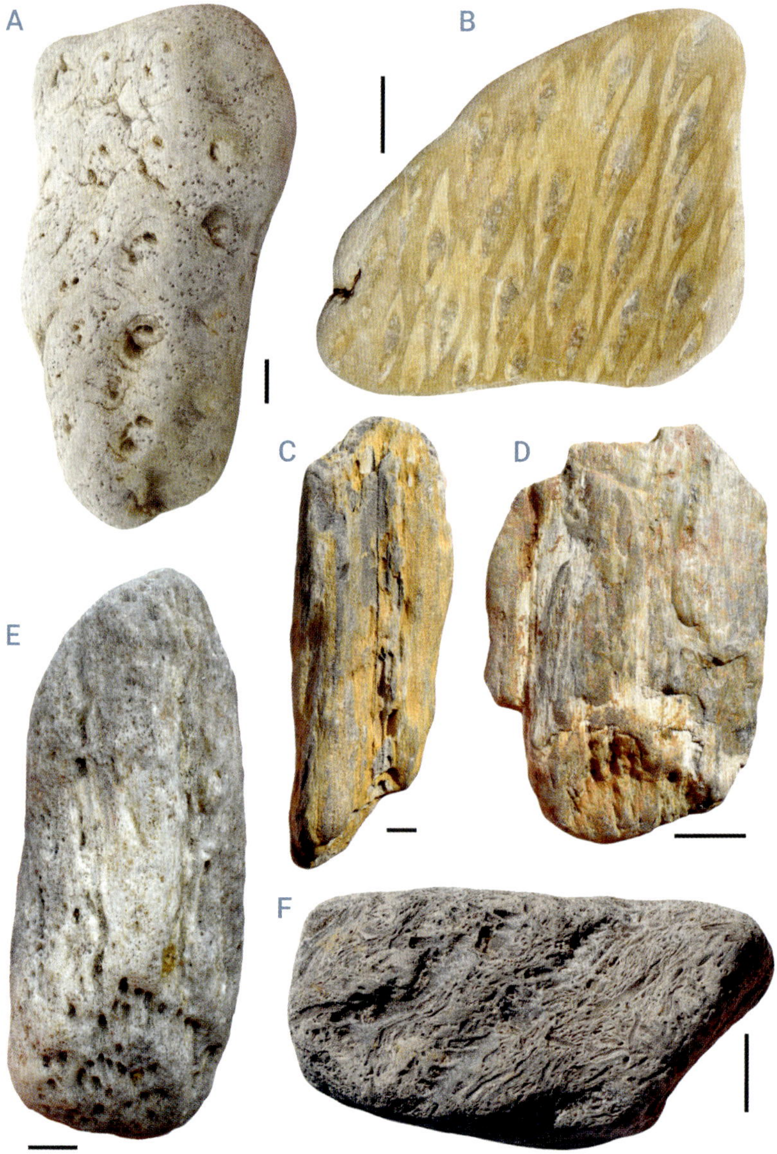

Figure 7.25

Examples of silicified wood. A, pithy core of a Cycad showing the phloem and xylem system, which allows the transport of water and nutrients up and down the plant. B, outer surface of a Cycad, with the diamond patterns representing the scars from dead leaves. C, D, sections of coniferous tree trunks. E, section of ?tree fern showing the vascular system. F, section of *Tempskya* trunk, which was actually a collection of stems surrounded by adventitious roots. F, collection of Huxley Hunt. Scale equals 10 mm.

Silicified Wood

The origin of this wood remains unclear but silicified wood comes in all colours and types (*Fig. 7.25*). Here are the most common types of silicified wood found.

- Cycad. The diamond structure of the outer surface of cycad tree trunks are the scars from fallen leaves as the tree grew. The inner surface shows the phloem and xylem system, which allows the transport of water and nutrients up and down the plant stem. The second chunk of cycad is the pith. This is the centre of the trunk, which is a soft region used for nutrient storage.
- *Tempskya*. This has a very ornate structure, with small pores, usually infilled with sparky quartz. These are easy to identify and are extremely pretty fossils to find! *Tempskya* is comparable to modern tree ferns.
- Indeterminate chunks. These are beach rolled and show the 'classic' woody structure.

Hanover Point Pine Raft

At Hanover Point, about 100 metres from the headland, is the so-called fossil forest or pine raft. This is an accumulation of large logs which were caught in a flood and formed a log jam on the inside of a river meander (*Fig. 7.26*). These have been permineralized and now form large boulders on the foreshore, often covered in seaweed and barnacles and the details of the wood are not always clear. The whole trees have since broken into segments, forming large cylindrical boulders in lines, the largest segment being over 1 m across and 3 m long.

Amber

On rare occasions, amber can be found in the plant debris beds. Unlike the honey-coloured Baltic amber seen in most fossil shops, Wessex Formation amber is darker, and may even have reddish hues (*Fig. 7.24*). Generally, it ranges from red to almost black and can be quite opaque. Wessex Formation amber is very brittle, breaking apart very easily under finger pressure. The amber found on the island has yielded many inclusions, ranging from spiders to midges.

Figure 7.26

Hanover Point Pine Raft.

A, broken section of a log showing the wood texture on the surface.
B, large log in situ.
C, large broken logs.

Trace Fossils

The Wessex Formation has numerous trace fossils throughout the sequence, usually seen in cliff or foreshore sections, but in most cases, it is impossible to collect them. Photographs with a scale are the best way to document trace fossil finds. For dinosaur footprints, *see page 177*.

In the sandstone units, a variety of different burrow structures

Figure 7.27

Trace fossils.

A, Bioturbated sandstone horizon in the Sudmoor Point sandstone, showing many vertical burrows. B, infilled burrows, likely made by very small mammals, reptiles or insects. C, B, oxidisation structures around plant roots.

are seen, especially on fresh surfaces. These can be simple burrows, or burrows with U-shaped structures inside called menisci, which is the backfill made by the organism as it packs the sediment behind it as it burrows (*Fig. 7.27*).

In some of the paleosols, rhizoliths, the traces of plant roots, can be seen as dark brown lines. And on the clays above on foreshore deposits, circles can be seen, usually a distinct colour to the surrounding clay with a halo. These are where trunks of trees and cycads were growing.

The study of trace fossils is called ichnology. Ichnologists, those who study trace fossils, use a naming system similar to that used for animals and plants, with ichnofamilies, ichnogenera, and ichnospecies. Consequently, almost every type of burrow, footprint or root trace has its own scientific name. The 3 toed dinosaur footprints made by ornithopod dinosaurs, like *Iguanodon*, are called *Iguanodontipus*. Many trace fossils are unique to certain environments, so knowing the common and more easily identifiable traces allows you to identify which sedimentary environment the rocks were deposited in.

Coprolites (fossilised faeces) are also classed as trace fossils (*Fig. 7.28*). Coprolites are very common within the Wessex Formation, often appearing exactly how you would imagine – sausage-shaped with distinct folds or striations. Coprolites can be

Figure 7.28
Wessex Formation coprolites.
B, D, collection of Huxley Hunt. Scale equals 10 mm.

incredibly useful for palaeontologists, often preserving the animals' last meal (pieces of bone, fish scales etc.), as well as traces left by insects that have burrowed into it. The shape and size can also indicate which animal it may have come from.

7.2. Fauna of the Vectis Formation lagoons

Although the Vectis exposures are rarely free of mudflows and landslides, the fossils are common on the foreshores at the base of the formation. The most productive fossiling comes from examining the fallen limestone slabs for visible remains and scouring the foreshores.

Gutter Casts

These are hard fine sandstone and siltstone infills of small, sinuous eroded channels in the lagoon floor muds of the Vectis lagoons,

Figure 7.29

Vectis Formation gutter casts.

A, fresh from the cliff showing original bedded structure and rippled shape. B, gutter cast full of bones. C, rolled pebble of gutter cast. C, collection of Huxley Hunt. Scale equals 10 mm.

likely caused by storm events. The eroded channels acted as a catchment for vertebrates and occasionally plant remains, which accumulated over time. Most of the preserved remains are fish bones (*Fig. 7.29*), but pterosaur teeth and bones, crocodile teeth and shark fin spines and teeth can also be found caught within these gutter casts. Unfortunately, it is very difficult to extract the small fossils as the rock is cemented by siderite, which does not break down in acetic acid.

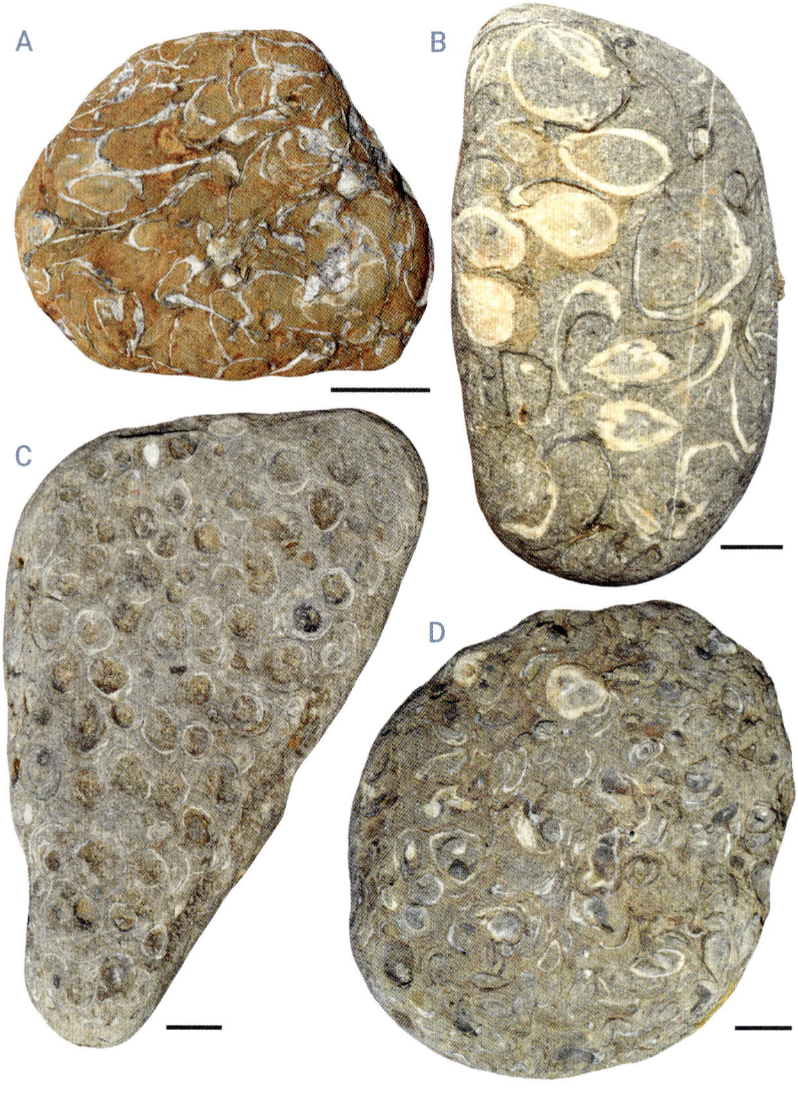

Figure 7.30

Rolled slabs of *Filosina* bed and oyster bed.

A, oyster bed. B, coquina. C, natural accumulation with shells orientated the same way up. D, coquina with some natural sorting of shells. Scale equals 10 mm.

Bivalves

The Vectis Formation yields a high diversity of bivalves, with external moulds and casts occurring in sideritic concretions, 2D compressed bivalves occur within the clay layers. However, the most common bivalves form limestone layers.

Compton and Yaverland beaches are often littered with large flat slabs from the Vectis Formation, usually concentrated at the base of the cliff in front of larger landslides. These are thin limestone horizons from within the upper part of the Vectis Formation, mostly made of an accumulation of a small bivalve called *Filosina* or a small silvery-grey oyster called *Liostrea* (*Fig. 7.30*). There are a few types of 'shelly limestone'. The most commonly occurring is the coquina, a storm event bed where all the shells are jumbled and broken. Some of the shells still have both valves closed, indicating that they were most likely buried alive. Other shell beds are natural accumulation over time, with all the shells laid the same way up. This winnowing disrupted the bivalves and moved them around until they were in a current stable position. The stacking of the shells can be seen on the side of the slabs.

These beds also yield a large amount of vertebrate remains, normally quite well preserved but isolated. These are easy to spot, as the black colour and sometimes shiny appearance stand out against the dull grey limestone. The most common finds are shark fin spines, shark cephalic spines (head spines), enamel fish scales, small bones, and teeth including those of rare pterosaurs.

Gastropods

Gastropods are found within the shelly limestone blocks, as well as within some of the sideritic concretions in the lower part of the Vectis Formation. Very fragile gastropods can also be found by splitting the clay; however, these have usually been compressed and crumble apart quickly. The most prevalent gastropod is *Viviparous*, which commonly occurs within the shelly limestones (*Fig. 7.31*) and can also be found in filling the dinosaur footprints (*see page 182*).

Figure 7.31

Viviparus gastropods.

A, internal and external moulds within a limestone nodule. B, crushed shells in a clay bed. Scale equals 10 mm.

Boney fish

The Vectis Formation yields a high diversity of fish, most often found as disarticulate bones and teeth scattered on the limestone blocks or within the gutter casts (*Fig. 7.32*). Fish vertebrae are easy to spot, either hourglass or X-shaped objects or a circle depending on the angle of the cross-section. These can range from one or two millimetres across to several centimetres across. Skull elements are usually flat bones, with most being difficult to identify without expert knowledge.

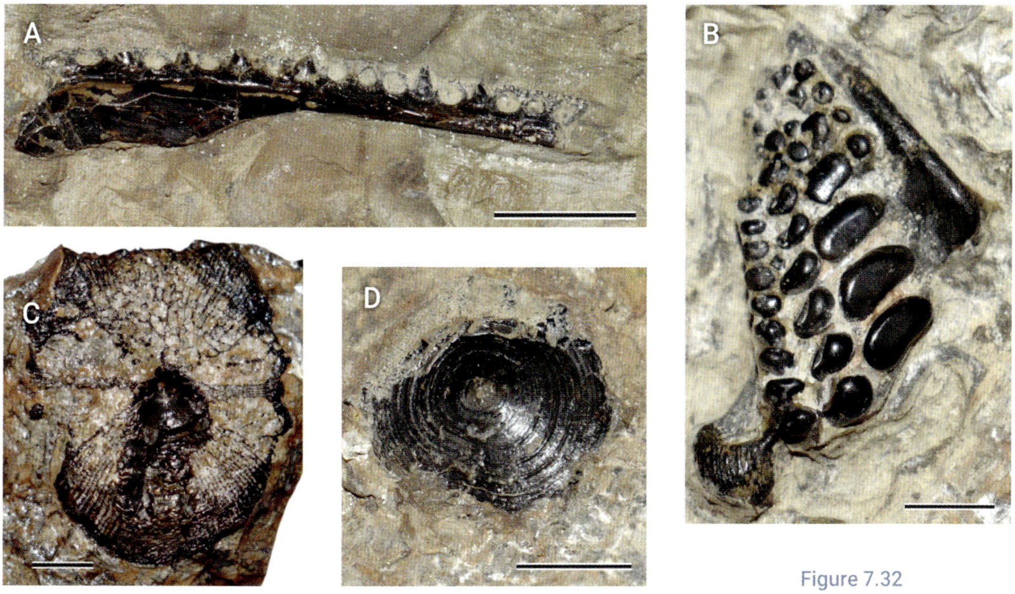

Figure 7.32

Fish remains from the *Filosina* beds. A, fish jaw. B, pycnodont jaw. C, worn fish vertebra showing the internal structure. D, fish vertebra. A, B, D, collection of Scott Cade. Scale equals 10 mm.

Pycnodont tooth palettes are very distinctive, possessing rows of short, shiny rounded teeth. Such teeth would be employed to crush hard-shelled prey such as molluscs and arthropods (*Fig. 7.32*).

Cartilaginous fish - Sharks

The shark found in the Vectis Formation is called *Egertonodus basanus*. This shark grew to around 2 metres long and processed two dorsal fins. In front of both dorsal fins were bony spines, called shark fin spines (*Fig. 7.33*). These are long, tapered, and slightly curved, with ridges extending along their length. They also have a row of tooth-like denticles (bumps) on their posterior margin. *Egertonodus* also possesses spines on the top of the head called cephalic spines, or head hooks. These are short and hook-like with a barb at the end. The base is usually T-shaped.

Turtles

Sections of turtle shell are a common occurrence within the shelly limestone slabs, easily identified as thin, flat pieces of bone. Occasionally, more complete carapaces and plastrons are found, usually slightly flattened due to compaction (*Fig. 7.34*). There are two named turtles from the Vectis Formation, *Plesiochelys* and *Brodiechelys*.

Figure 7.33

Shark remains from the *Filosina* beds.

A, *Egertonodus* cephalic hook. B, C, shark fin spines showing variation in ornamentation. D, *Hybodus* tooth. A, collection of Roy Smith, photo courtesy of Lucy Smith. C, collection of Scott Cade. A - C, scale equals 10 mm. D, scale equals 1 mm.

Crocodiles

The crocodiles of the Vectis Formation are poorly known, with only one named species described, *Hylaeochampsa vectiana*. This is likely due to the lack of complete remains. However, isolated scutes and teeth can be found within the limestones and clays.

Figure 7.34

Turtle carapace from the Vectis Formation. Collection of Jack Wonfor. Scale equals 10 mm.

Marine Reptiles

Marine reptile remains are exceedingly rare in the Vectis Formation, but occasionally turn up. The only named marine reptile is the plesiosaur *Vectocleidus pastorum*. Plesiosaurs, usually a marine species, began venturing into freshwater environments in the Early Cretaceous. In particular, a family of plesiosaurs called Leptocleidiae, occur more frequently in lagoonal and even occasionally riverine settings. They were small plesiosaurs with short necks. To date, there have been 12 occurrences of these plesiosaurs in freshwater habitats across the globe, including Australia, Morocco, the US and Germany.

Plesiosaur remains are easy to identify. Their vertebrae are circular, short in length with a slightly dished surface at the front and back. If it is a mature individual, the neural arch will have fused to the centrum. The teeth are slender, slightly curved with no serrations and a circular cross section. The humerus and femur are flattened, with a circular end broadening to a wider end. The podial bones, which make up the portion of the paddle, are hourglass-shaped.

Pterosaurs

Pterosaurs are flying reptiles, which dominated the skies between 215 million years ago to 66 million years ago, becoming extinct with the dinosaurs at the end Cretaceous mass extinction. There are two named pterosaurs from the Vectis Formation: *Istiodactylus latidens* and *Lonchodectes sagittirostris*. There is also an unidentified ornithocheirid, but so far only its teeth have been found. Pterosaur remains are rare compared to other fossils, however, they do turn up from time to time, usually occurring within the shelly limestone slabs or pyrite nodules.

Pterosaurs have a highly specialised skeleton modified for flight, with hollow (pneumatic) bones to reduce as much weight as possible. All the bones are slender, with paper-thin bone walls, with delicate, thin struts inside the bone which add strength and resist torsional stresses (*Fig. 7.35*).

Pterosaur teeth occur fairly frequently, usually occurring as isolated crowns in the shelly limestone slabs. The most common teeth are those of ornithocheirid pterosaurs, which are long and slenderly curved (*Fig. 7.36*). They are easily distinguished from other reptile teeth by having a U-shaped boundary between the dentine and enamel layers. The enamel layer usually has a wrinkly texture on the surface. The front teeth have a rounder cross section and are more curved, while the teeth further back in the jaw are straighter

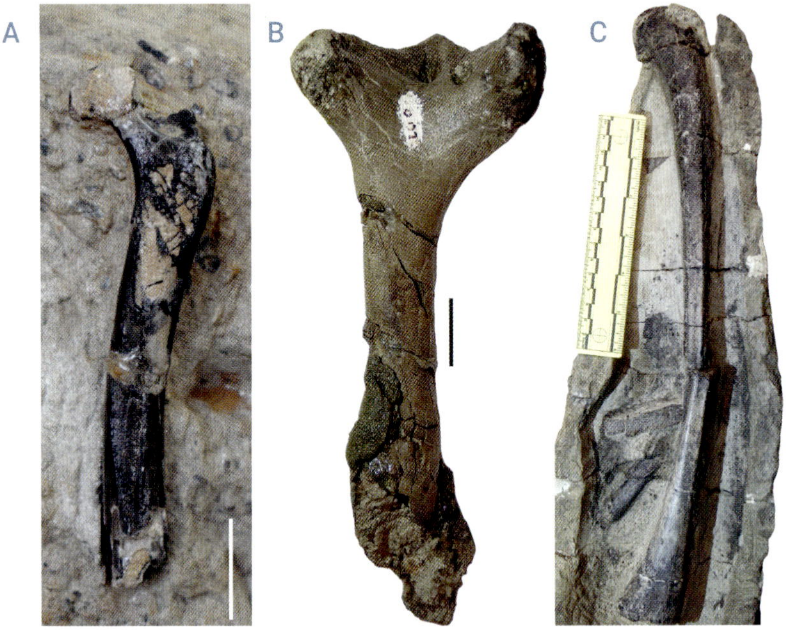

Figure 7.35

Right: A, pterosaur femur showing hollow internal structure supported by thin struts of bone. B, pterosaur humerus. C, pterosaur wing bone. A, collection of Maria Yule and Luke Ferguson. B, C, collection of Dinosaur Expeditions C.I.C. Scale equals 10 mm.

Figure 7.36

Below: A, Pterosaur tooth in a *Filosina* block with B, a diagram showing pterosaur tooth anatomy. Scale equals 10 mm.

with a more oval cross-section and triangular outline in lateral view. *Istiodactylus* teeth are short, triangular, and compressed, and exceedingly rare.

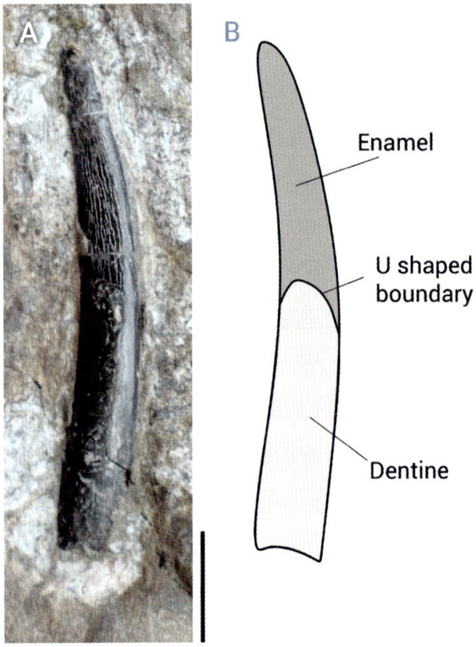

Ostracods

When splitting fallen blocks of clay, you may come across surfaces covered in tiny little white baked beans. These are in fact ostracods, a type of small crustacean, sometimes known as a seed shrimp (*Fig. 7.37*). They consist of two clam-like valves (shells), which open and close. Allowing numerous arms and legs to protrude for locomotion and feeding. They are usually around 1-2 mm in length and are part of the benthos (community of animals that live on or in the sand/mud), although some species can be planktonic (drifting in the water column).

Figure 7.37

A, section of the ostracod bed at Compton. B, close-up showing a dense mass of ostracod shells.
C, an extant ostracod.

Reworked Fossils

There are several reworked fossils to be found as well, mainly within the shelly limestone slabs. These are fossils that were preserved in older Jurassic rocks, 161–145 million years old (Oxfordian to Tithonian) rocks, which had been eroded out of the clay and redeposited into the Vectis lagoons. You can commonly find segments of ammonite, echinoid spines and silicified oysters.

Trace Fossils

Trace fossils within the Vectis Formation are common on the *Filosina* slabs, with pill-shaped structures called *Diplocraterion*, and surfaces covered in small, thread-shaped burrow structures (*Fig. 7.38*).

Casts of desiccation cracks are also a common occurrence (*Fig. 7.38*). These are evidence of the lagoon drying up, and the mud cracking. Overlying mud then covered the surface, infilling the cracks, and producing what we see today. For dinosaur footprints in the Vectis Formation, *see page 182*.

7.3. Dinosaur Footprints and Foot casts

Footprints are a window into the dinosaurs' world, capturing information such as behaviours, how the animals moved, the length of their strides and even their walking speed. Such information cannot be deduced from their skeletal remains. The dinosaur footprints on the island are known worldwide, with the foot casts

Figure 7.38

Trace fossils of the Vectis Formation.

A, ?*Planolites* burrows. B, the underside of a gutter cast with scour marks and burrows. C, the underside of a block of *Filosina* bed with ?*Lockeia* bivalve trace fossils and small thread-like burrows. Scale equals 10 mm.

at Hanover Point visible all year round. This allows you to see evidence of the island's dinosaurs, even when the beach is lacking in fossil finds.

Hanover Point Foot Casts

The Hanover Point dinosaur foot casts are probably the most famous trace fossils on the island, known to science since 1853, there are now over 250 foot casts on the beaches around Hanover

Figure 7.39

Dinosaur foot cast boulder field at Hanover Point.

Point, with more appearing every year (*Fig. 7.39*). The usually 3-toed casts are mostly referred to *Iguanodon*, however many different types of ornithopods would have made the original footprints. There are also theropod, sauropod and ankylosaur foot casts on the beach, however, these are a lot rarer than the ornithopod foot casts (*Fig. 7.40*).

The Hanover Point foot casts were made by dinosaurs walking through deep mud, probably to a water source, leaving deeply impressed footprints. This mud then hardened. The Hanover Point Sandstone is a crevasse-splay sandstone, that forms when a river

Figure 7.40

Hanover Point foot casts

A, iguanodontid foot cast with a handprint. B, ?*Polacanthus* foot cast. C, iguanodontid foot cast with a small foot cast on top. D, small theropod foot cast. E, iguanodontid foot cast showing claw drag marks. F, iguanodontid foot cast in situ.

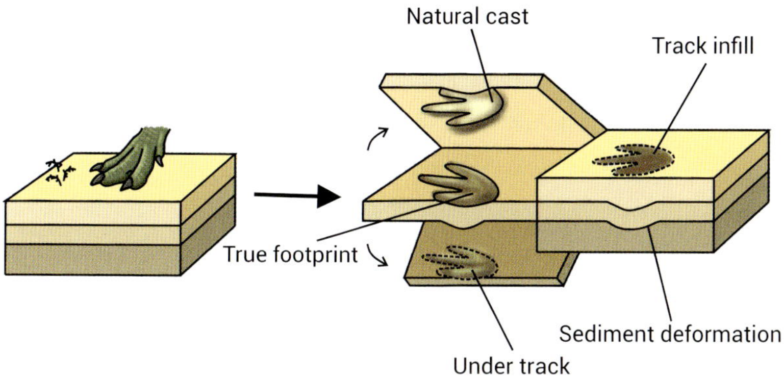

Figure 7.41
Diagram showing the Formation of footprints and foot casts.

bursts its banks and flows onto the floodplain. The sand suspended in the water falls out of suspension as the water velocity drops and infills the footprints. Over time and with continued burial, the sand turns to a hard sandstone. Today this section of the floodplain is exposed in the cliff and on the foreshore, where the sea washes away the soft clay surrounding the footprint, leaving a perfect cast (*Fig. 7.41*). They can be seen on the underside of the sandstone where it forms a prominent ledge extending out to sea at the point itself. Today the constant erosion from the seas and rain removes the clay beneath the sandstone, triggering cliff falls and leaving the hard foot casts washed out nice and clean on the beach.

In the past, unscrupulous collectors have used sack barrows or driven their vehicles down Brook slipway onto the beach to remove the foot casts. However, since awareness has been raised, there are now hundreds of foot casts on the beach for everyone to enjoy.

Other footprints

Besides the foot casts at Hanover Point, many other vertebrate trace fossils occur in the Wessex and Vectis formations. There are many footprints within the foreshore clays, which are exposed on every low tide (*Fig. 7.42*). Due to the softer nature of the clays, they tend to erode quickly, therefore their appearance is often only brief.

The best-known trackway is at Hanover Point on a sandstone on the foreshore that can only be accessed on a low spring tide (*Fig. 7.43*). These are a set of footprints on the top of a sandstone, with another set slowly appearing on the surface below. This trackway was first discovered in the 1800's and has survived for more than a century. Unfortunately, it was vandalised in the 90's, when a selfish collector removed one of the footprints with a rock saw.

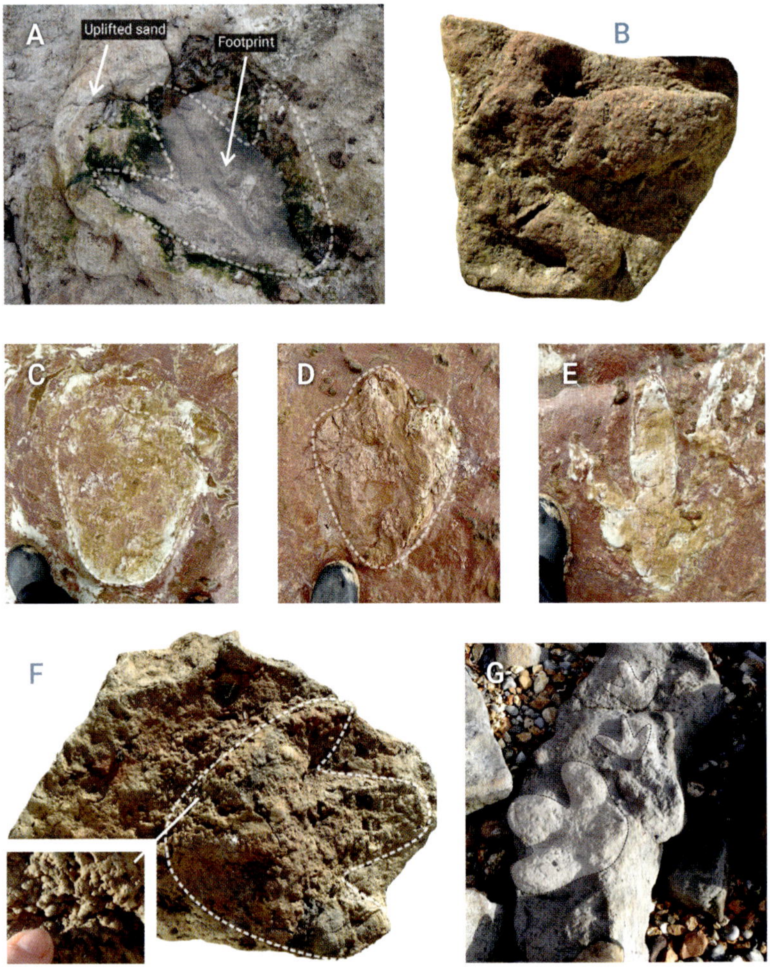

Figure 7.42

Other examples of dinosaur footprints. A, footprint on the top surface of the Hanover Point Sandstone, with disturbed sediment around one edge. B, theropod foot cast in the *Filosina* bed from the Vectis Formation. C, sauropod footprint in the red beds near Sudmoor Point. D, iguanodontid footprint in the red beds near Sudmoor Point. E, theropod footprint in the red beds near Sudmoor Point. F, iguanodontid foot cast in the *Filosina* beds, with *Viviparous* gastropods infilling the footprint. G, shallow dinosaur foot casts at Hanover Point, showing a small iguanodontid trackway. B, F, collection of Dick Downes

Take great care when viewing the trackway as the rocks are often extremely slippery, often covered in a thin layer of silt and algae, especially in summer months.

There are also some well-defined footprints on the top of the Hanover Point sandstone, showing the sand being displaced as the dinosaur walked across the riverbank (*Fig. 7.42A*). Footprints can be seen in many other sandstones through the Wessex Formation, either as casts at the base (in situ and loose on the beach) or as a cross-section in the cliff showing the sediment displacement.

Small footprints can also be found on the underside of the shelly

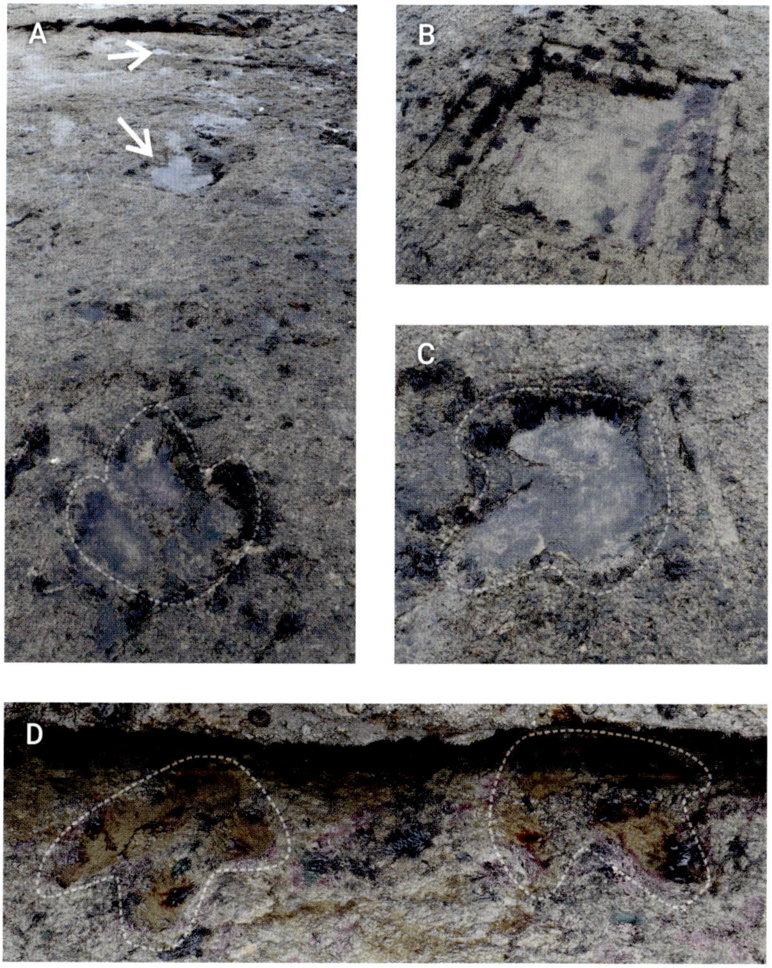

Figure 7.43

The Hanover Point Trackway, showing the A, sequence of footprints heading north. B, large square hole where a selfish collector used an angle grinder to remove one of the footprints. C, newly exposed footprints on the sandstone layer below.

limestone blocks from the Vectis Formation. These are often infilled with freshwater gastropods (snails), *Viviparus*, and can be found at Compton Bay and Yaverland (*Fig. 7.42B,F*). These dinosaurs would have been crossing the Vectis lagoons, indicating that, at least for some periods, the water levels were low enough to allow small dinosaurs to wade in the waters.

What dinosaur made the track?

There are several distinct types of dinosaur footprints, indicated by the different shapes of footprint left behind (*Fig. 7.44*). The most common footprints present on the island are those made

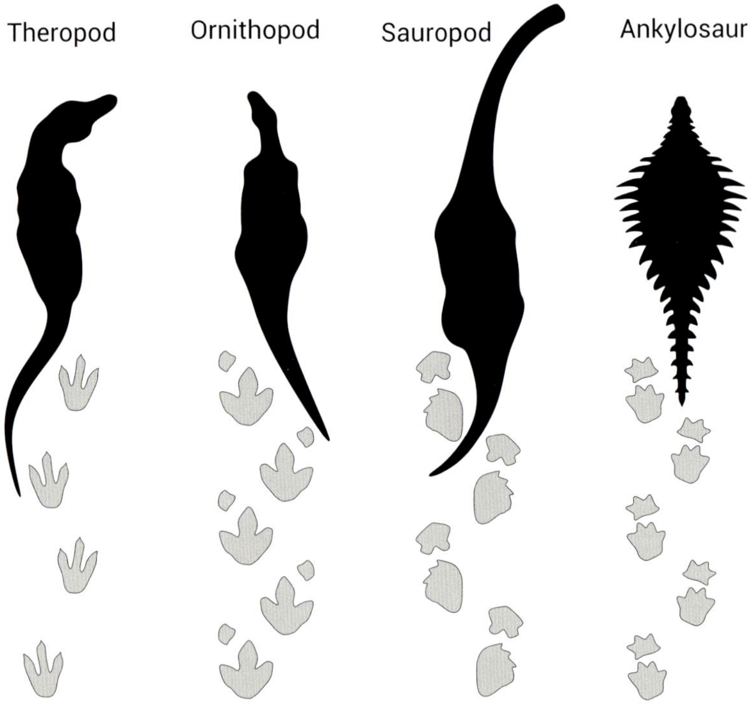

Figure 7.44

Diagram showing dinosaur footprint shape.

by ornithopod dinosaurs such as *Iguanodon*, *Mantellisaurus* and *Brighstoneus*. These three-toed footprints, with more robust, blunt toes, have rounded claws, and wider spread toes. Theropod (carnivorous) dinosaurs also made three-toed footprints, but the toes tend to be more elongate, with sharply pointed claw impressions. Sauropod footprints are generally large, and rounded, with five blunt toes that curve around the side of the footprint. The handprints are very different in sauropod dinosaurs, being more crescent-shaped. Ankylosaur footprints, such as *Polacanthus* and *Vectipelta*, are extremely rare compared to the other dinosaurs. These are smaller, 4-toed prints, with an elongate foot and the toes at the front of the foot. No one has yet found the footprints of pterosaurs on the island, but it is surely just a matter of time before they are discovered.

7.4. Fauna of the Lower Greensand, Gault Clay, Upper Greensand and Chalk seas

The Lower Greensand, Gault Clay, Upper Greensand and Chalk were all deposited in Cretaceous seas during a major sea level rise

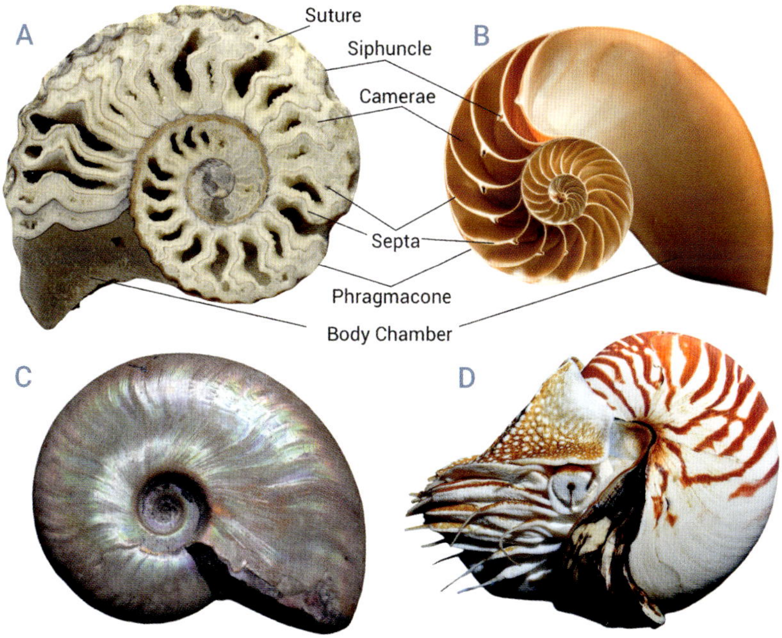

Figure 7.45

Diagram showing A, ammonite and B, Nautilus morphology. C, a *Cleoniceras* from Madagascar with the shell showing pearlescent shell. Although this ammonite is from Madagascar, the same ammonite can be found in the Carstone on the island. D, an extant Nautilus.

starting in the latter part of the Early Cretaceous. So, in this section, to avoid repetition, the faunas of all four formations have been combined into their respective groups, with illustrations starting from oldest through to youngest.

Ammonites and Nautiloids

Ammonites are a type of mollusc called a cephalopod, the name meaning 'head foot'. Cephalopods include squid, cuttlefish, octopi, argonauts and nautili. Most ammonites possessed a hard, chambered spiral shell, like that of the living nautilus, with a presumed squid-like animal living inside (*Fig. 7.45*). Ammonites are rather enigmatic as no external soft tissue (such as tentacles) has been found preserved. Consequently, what an ammonite looked like when it was alive is a bit of a mystery. The lack of ammonites with soft tissue preservation is also a big question in palaeontology, as many examples of fossil squids, octopi, and cuttlefish, all with soft tissue preservation in the same deposits as ammonites. Yet the ammonites are only preserved as hard shells, so there must have been something strange going on with their physiology.

Parts of an ammonite

PHRAGMACONE. This is the chambered portion of the ammonite shell, which is divided by septa into camerae. This is the only biomineralized part of the ammonite, and typically the only part that is preserved in the fossil record.

CAMERAE OR CHAMBERS. As the ammonite grows, it adds larger chambers to the open end of the coil. The last and biggest chamber is called the body chamber and contains the body of the living animal.

SEPTA. Each chamber is separated by a thin wall called a septum (plural: septa). As the animal grows and moves forward into the new body chamber, it secretes aragonite behind it to create the septum.

SUTURES. The suture line is the boundary between two septa or chambers. During the ammonite's evolution, these chambers became increasingly complex. It is thought that the greater complexity of the suture pattern added strength to the shell, allowing the animal to withstand the pressure at 100s of metres' water depth.

SIPHUNCLE. All the chambers of the ammonite are connected by the siphuncle. This is a tube, once containing living tissue, running along the outer region of the phragmacone. The siphuncle allowed for the fluid or gas to be pumped in and out of the chambers to control buoyancy and allow for the ammonite to move up or down in the water column.

Ammonite shell

The ammonite's shell is made of aragonite, a polymorph of calcium carbonate. It has the same chemical composition as calcite, but a different crystalline structure. The shell itself is composed of 3 layers, an inner and outer layer of aragonite prisms, with a thicker middle layer of tiny tabular aragonite crystals. These layers produce the mother-of-pearl shine, like modern nautilus shells (*Fig. 7.45*).

Aragonite is less stable than calcite, and consequently, it is easily replaced by other minerals, often neomorphing into calcite or being completely replaced by pyrite. Therefore, finding the original aragonite shell of an ammonite is a rare occurrence. On the island, the most common ammonites have either been replaced with calcite, or the chambers have been infilled with sediment, pyrite or calcite, leaving only internal moulds of the ammonite's phragmacone.

Heteromorph ammonites

Most ammonites grew in a regular plani-spiral. However, at several points in ammonite evolution, the spiral has started to uncoil, resulting in a variety of bizarre shapes, from just a simple loose coil to intertwining coils, completely straight and giant paper clip-shaped ammonites. Several of these heteromorph ammonites occur on the Isle of Wight and are eagerly sought after by collectors.

These complex shaped shells would have made the animals poor swimmers compared to their more normally coiled cousins. The lack of a siphonic notch, meaning it's siphon was very small and likely could not produce the same amount of jet propulsion, further suggests that they were poor swimmers.

It is thought these forms would have been 'hanging' in the water column, collecting small food particles on long tentacles, much like jellyfish do. Alternately, some heteromorphs may have lived on the ocean floor, feeding on carcasses or slow moving or sessile animals. The Lower Greensand and Chalk yield heteromorph ammonites, with the genera *Scaphites*, *Tropaeum* and *Turrilites* occurring most commonly.

Ammonites on the Isle of Wight

The marine strata on the island produce hundreds of different ammonites, with preservation ranging from internal moulds to examples with the original shell and mother-of-pearl lustre. The glauconitic marl especially yields a high diversity of ammonites, many with a high degree of variation between individuals of the same species. This makes it extremely difficult to distinguish between different species, leading to many of the ammonites having the species name 'variens'.

The coastline from Atherfield Point to Rocken End is incredibly rich in ammonites (*Figs. 7.46 - 7.56*). However, some are so large, namely *Tropaeum* and *Epicheloniceras*, that retrieving them requires a huge team effort.

Figure 7.46

Lower Greensand ammonites.

A, *Deshayesites forbesi*. Upper lobster beds, Atherfield Clay Formation, Lower Greensand, Chale Bay. B, cut and polished *Deshayesites forbesi* from the Upper lobster beds showing internal morphology. C, *Deshayesites saxbyi*. Lower crackers nodules, Crackers member, Atherfield Clay Formation, Chale Bay. D, *Deshayesites forbesi*. Upper Lobster beds, Atherfield Clay Formation, Chale Bay. E, pyrite *Deshayesites forbesi*. Upper Lobster beds, Atherfield Clay Formation, Chale Bay. F, *Deshayesites planus*. Upper Lobster bed, Atherfield Clay Formation, Chale Bay. G, *Deshayesites normani*. Upper lobster beds, Atherfield Clay Formation, Chale. A, B, C, D, collection of Glenn Martin. E, F, G, collection of Jack Wonfor. Scale equals 10 mm.

Cat paws are individual ammonite chamber steinkerns that occur commonly at Shanklin and Luccombe. These are individual chambers of the ammonite that have been infilled by minerals, often by black or very dark brown francolite (a form of calcium phosphate) (*Fig. 7.52F*).

Nautiloids

The nautilus is a type of pelagic (living in the open sea) cephalopod and is the only extant (still living) cephalopod with an external planispiral shell, due to surviving the end-Cretaceous mass extinction event, while all species of ammonite became extinct. Whilst the nautilus looks superficially like an ammonite, both having a chambered spiral shell, there are major differences that

Figure 7.47

Lower Greensand ammonites.

A, *Cheloniceras crassum*. Group V Scaphites bed, Ferruginous Sands Formation, Lower Greensand Chale Bay. B, *Deshayesites forbesi*. Lower crackers nodules, Crackers member, Atherfield Clay Formation, Chale Bay. C, *Deshayesites forbesi*. Upper lobster beds, Atherfield Clay Formation, Chale Bay. D, *Deshayesites vectensis*. Group V Scaphites bed, Ferruginous sands Formation, Lower Greensand, Chale Bay. E, *Deshayesites grandis*. Group V Scaphites bed, Ferruginous sands Formation, Lower Greensand, Chale Bay. F, *Deshayesites forbesi*. Lower lobster bed, Atherfield Clay Formation, Chale Bay. G, *Deshayesites forbesi*. Lower crackers nodules, Crackers member, Atherfield Clay Formation, Chale Bay. A, E, collection of Glenn Martin. B, C, D, F, G, collection of Jack Wonfor. Scale equals 10 mm.

make it possible to tell apart. Mainly, nautili have simple sutures, usually a curved line and a siphuncle that extends along the centre of the whorl (*Fig. 7.45*).

Nautili are a relatively rare find on the island, with the most common being *Eutrephoceras*, *Cymatoceras* and *Heminautilus* (*Fig. 7.57*).

Figure 7.48

Australiceras gigas from the Group V Scaphites Bed, Lower Greensand Group of Whale Chine showing variation within the species. A, collection of Jack Wonfor. B, C, D, collection of Glenn Martin. Scale equals 10 mm.

Bivalves

Bivalves are a type of mollusc and form a remarkably diverse and common group in both salt and freshwater environments. Almost all bivalves are filter feeders, sucking water into specialised organs which filter food particles from the water. Bivalves consist of paired valves (the shell), which are held together by interlocking processes called a hinge. Internally, there are two muscles (anterior and posterior adductor muscles), which connect the two valves and are

Figure 7.49

Lower Greensand Ammonites.

A, *Tropaeum bowerbanki*. Group VI Lower Crioceras bed, Ferruginous Sands Formation, Whale Chine. B, *Tropaeum bowerbanki*. Group VI Lower Crioceras Bed, Ferruginous Sands Formation, Whale Chine. C, *Cheloniceras crassum*. Group V Scaphites bed, Ferruginous sands Formation, Whale Chine. D, *Cheloniceras crassum*. Group V Scaphites bed, Ferruginous sands Formation, Whale Chine. E, *Dufrenoya transitoria*. Group VI Lower Crioceras bed, Ferruginous Sands Formation, Whale Chine. F, *Epancyloceras* sp.. Group VI Lower Crioceras bed, Ferruginous sands Formation, Whale Chine. A, B, collection of Jack Wonfor. C, D, E, F, collection of Glenn Martin. Scale equals 50 mm.

employed to open and close the shells. These adductor muscles leave attachment scars on the inside of the valves, usually one at either end, although in pectens and oysters there is often one large conspicuous scar in the middle (*Fig. 7.58*). The placement and shape of these scars are often used to identify species. Many bivalve shells are made from a form of calcium carbonate called calcite, which preserves well. Others can be made from aragonite, which is often replaced with

Figure 7.50

Lower Greensand Ammonites.

A, *Epicheloniceras tschernyschewi*. Group IX, Ferruginous Sands Formation, Walpen Chine. B, *Roloboceras hambrovi*. Lower crackers nodules, Crackers member, Atherfield Clay Formation, Chale Bay. C, *Roloboceras hambrovi*. Lower lobster bed, Atherfield Clay Formation, Chale Bay. D, *Roloboceras hambrovi*. Bed 10, Lower lobster bed, Atherfield Clay Formation, Chale Bay. E, *Epicheloniceras* sp.. Group IX, Ferruginous Sands Formation, Walpen Chine. F, *Deshayesites saxbyi*. Lower crackers nodules, Crackers member, Atherfield Clay Formation, Chale Bay. A, B, collection of Glenn Martin. C, D, E, collection of Jack Wonfor. Scale equals 10 mm.

calcite. Others, such as mussels, have both calcite and aragonite. However, bivalve shells can be dissolved by acidic waters, leading to just the internal and external moulds of the shells preserved.

The knowledge of modern bivalves aids our understanding of extinct species, as the different lifestyles influence the shape of the shell. There are 4 common modes of life for bivalves; infaunal (burrowing), nektonic (swimming), epifaunal; cementers, byssally attached and free lying (called recliners) and borers.

Figure 7.51

Lower Greensand Ammonites.

A, *Epicheloniceras tschernyschewi*. Group IX, Ferruginous Sands Formation, Walpen Chine. B, *Epicheloniceras tschernyschewi*. Group IX, Ferruginous Sands Formation, Walpen Chine. C, *Epicheloniceras debile*. Group IX, Ferruginous Sands Formation, Walpen Chine. D, *Epicheloniceras martinoides*. Group VIII Upper Crioceras bed, Ferruginous Sands Formation, Chale Bay. E, *Tonohamites decurrens*. Group V Scaphites Bed, Ferruginous Sands Formation, Chale Bay. F, *Epicheloniceras tschernyschewi*. Group IX, Ferruginous Sands Formation, Walpen Chine. A, C, F, collection of Glenn Martin. B, D, E, collection of Jack Wonfor. Scale equals 10 mm.

Burrowing bivalves live within the sediment, either just beneath the surface or deep within the substrate. A muscular foot is used to burrow. Burrowing bivalves typically have elongate shells, becoming more elongate the deeper the bivalve burrows. Burrowing species typically have a specialised tissue tube called a siphon to move water enabling the animal to respire, as well as filter out nutrients from the water. The siphon leaves a mark on the shell called the pallial sinus, which is greatly pronounced in deeper burrowers due

Figure 7.52

Gault Clay ammonites. A, *Hoplites dentatus*. Lower Gault, Gault Clay Formation, Selbourne Group, Steel Bay, Bonchurch. B, *Hoplites dentatus*. Lower Gault, Gault Clay Formation, Selbourne Group, Steel Bay, Bonchurch. C, *Hoplites dentatus*. Lower Gault, Gault Clay Formation, Selbourne Group, Steel Bay, Bonchurch. D, *Hoplites dentatus*. Lower Gault, Gault Clay Formation, Selbourne Group, Binnel Bay, St. Lawrence. E, *Beaudanticeras* sp.. Lower Gault, Gault Clay Formation, Selbourne Group, Steel Bay, Bonchurch. F, Steinkern of *Hoplites* sp.. A, B, C, E, F, G, collection of Jack Wonfor. D, collection of Theo Vickers. Scale equals 10 mm.

to having a longer siphon to reach the top of the sediment. Modern cockles are an example of a shallow burrowing species, and razor clams are a deep burrowing species.

Nektonic bivalves, such as scallops, use a single and powerful adductor muscle that expands and contracts to forcefully expel water, allowing the bivalve to move across the seafloor. These bivalves do not swim freely all the time, instead reclining on the seafloor while filter-feeding, but use the sudden bursts to escape predators. Nektonic bivalves have one flattened valve and one convex valve, with 'ears' around the umbo, which act to direct water currents.

Epifaunal bivalves, live above the sediment but cannot move. These can live in a variety of different ways. Cementing bivalves, such as oysters, start off as larvae which then attach to a hard substrate on the seafloor. They secrete a calcareous cement, fusing

Figure 7.53

Gault ammonites.

A, *Euhoplites truncatus*. Lower Gault, Gault Clay Formation, Selbourne group, Steel Bay, Bonchurch. B, D, *Hoplites maritimus*. Lower Gault, Gault Clay Formation, Selbourne Group, Steel Bay, Bonchurch. C, E, *Anahoplites osmingtonensis*. Lower Gault, Gault Clay Formation, Selbourne Group, Steel Bay, Bonchurch. F, G, *Lyelliceras* sp.. Lower Gault, Gault Clay Formation, Selbourne Group, Steel Bay, Bonchurch. A, collection of Theo Vickers. B-G, collection of Jack Wonfor. Scale equals 10 mm.

their left valve to the substrate. Oysters will take on the shape of their substrate, and there are many instances of ammonite impressions that have been found on fossil oysters. As cemented bivalves are unable to move away from predators or danger, cementing bivalves often have incredibly thick shells for protection.

Byssally attached bivalves, such as mussels, produce strong and flexible proteinaceous threads, called a byssus, to attach themselves to a solid surface. Often, juvenile bivalves use a byssus for stability. These bivalves will also have large adductor muscles which enable them to hold both valves shut for protection, either against predators, high energy environments or when the animal is exposed during a low tide.

Free-lying bivalves or recliners, like the well-known oyster *Gryphaea* (Devil's toenail), do not cement themselves to the seafloor, but simply rest on it. The left valve is convex, which settles onto the

Figure 7.54

Upper Greensand ammonites.

A, *Mortoniceras inflatum*. Sandstone and rag beds, Upper Greensand Formation, Selbourne Group, Woody Bay, St. Lawrence. B, *Callihoplites pulcher*. Sandstone and rag beds, Upper Greensand Formation, Selbourne Group, Binnel Bay, St. Lawrence. C, *Mortoniceras inflatum*. Sandstone and rag beds, Upper Greensand Formation, Selbourne Group, St. Lawrence. D, *Callihoplites fuscus*. Sandstone and rag beds, Upper Greensand Formation, Selbourne Group, Steel Bay, Bonchurch. A, B, C, collection of Jack Wonfor. D, collection of Scott Cade. Scale equals 10 mm.

surface of the sediment, preventing it from sinking and allowing the water to flow around it. It's thought that these live in low-energy environments, so they weren't in danger of being washed away by a strong current.

Boring bivalves are a group that create a burrow in wood or solid rock using several different mechanisms. Some bivalves, like the modern Date Mussel (*Lithophaga*), bore into solid rock. They secrete an acid, which etches the rock. Once softened, the bivalve uses its rough shell to mechanically file a hole. Other boring bivalves use abrasion of their shells to bore into softer rocks, such as the modern Piddock (also known as angelwings), which lives in soft sandstones and clays.

Wood boring bivalves, such as shipworm *Teredo*, are a group of bivalves that use their shells as a file to bore holes into marine driftwood. The traces of these are commonly found in pieces of fossil driftwood from the Lower Greensand Group at Shanklin. The efficiency of wood boring was a big problem for hundreds of years

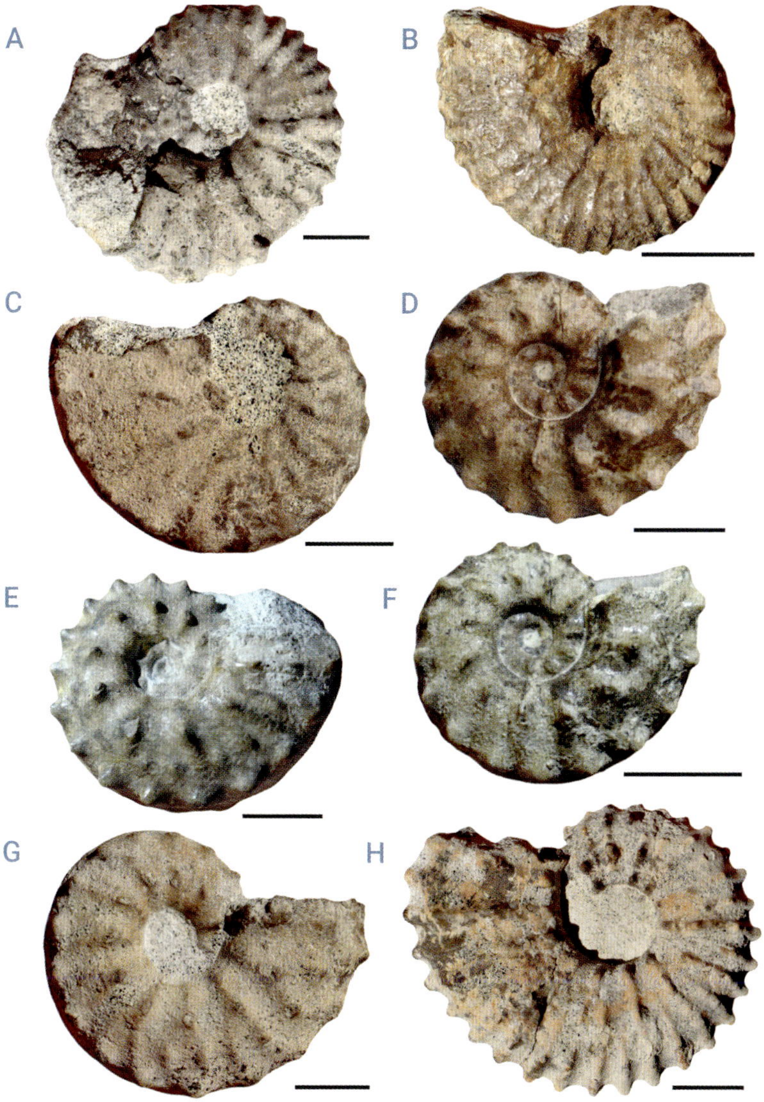

Figure 7.55

Chalk ammonites.

West Melbury Marly Chalk Formation.

A, *Mantelliceras mantelli*. Glauconitic marl member, Grey Chalk sub-group, Woody Bay, St. Lawrence. B, *Mantelliceras mantelli*. Glauconitic marl member, Grey Chalk sub-group, Woody Bay, St. Lawrence. C, *Schloenbachia varians*. Glauconitic marl member, Grey Chalk sub-group, Woody Bay, St. Lawrence. D, *Schloenbachia varians subtuberculata*. Glauconitic marl member, Grey Chalk sub-group, Rocken end, Niton. E, *Schloenbachia varians subtuberculata*. Glauconitic marl member, Grey Chalk sub-group, Woody Bay, St. Lawrence. F, *Schloenbachia varians subtuberculata*. Grey Chalk sub-group, Steel Bay, Bonchurch. G, *Schloenbachia varians*. Glauconitic marl member, Grey Chalk sub-group, Steel Bay, Bonchurch. H, *Mantelliceras mantelli*. Glauconitic marl member, Grey Chalk sub-group, Mount Bay, St. Lawrence. A, B, C, H, collection of Theo Vickers. D, E, F, G collection of Jack Wonfor. Scale equals 10 mm.

for wooden ships. Studies in the Baltic Sea showed 30 cm wide whole trees were destroyed within a year by shipworms!

The marine Cretaceous rocks on the island yield a wide diversity of bivalves, in the form of complete and articulated specimens to nodules full of internal and external moulds (*Figs. 7.59 – 7.64*).

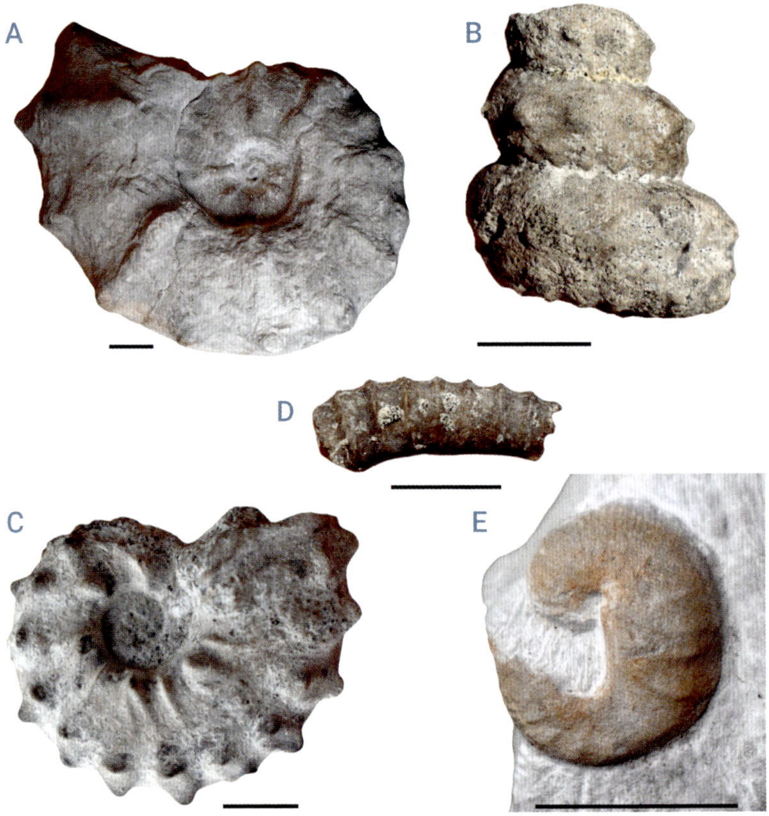

Figure 7.56

Chalk ammonites.

A, *Acanthoceras jukesbrownei*. Zig Zag Chalk Formation, Grey Chalk sub-group, Culver cliff. B, *Hypoturrilites* sp. Glauconitic marl member, West Melbury Marly Chalk Formation, Grey Chalk sub-group, Orchard Bay, St. Lawrence. C, *Hyphoplites curvatus curvatus*. Glauconitic Marl member, West Melbury Marly Chalk Formation, Grey Chalk sub-group, Rocken end, Niton. D, *Hamites* sp. Glauconitic marl member, West Melbury Marly Chalk Formation, Grey Chalk sub-Group, Ventnor. E, *Scaphites equalis*. Zig Zag Chalk Formation, Grey Chalk sub-group, Culver Cliff. A, C, E, collection of Jack Wonfor. B, D collection of Theo Vickers. Scale equals 10 mm.

Gastropods

Gastropods are a type of mollusc, commonly known as snails and slugs. Gastropods are very diverse and a highly successful group, inhabiting marine, freshwater and terrestrial environments with a global distribution. There are at least 15,000 extinct species of gastropod, and 85,000 extant species. Gastropods have a large muscular foot used for locomotion, with stalked eyes and tentacles for sensing their surroundings. Many have specialised rasp-like mouth parts, called a radula, consisting of lots of tiny teeth. Some gastropods have an operculum, which is a calcareous dish-like structure attached to the upper surface of the foot. When the gastropod retreats inside its shell, the operculum acts as a 'trapdoor', sealing the entrance to the shell for protection (*Fig. 7.65*).

Most gastropods consist of an aragonite spiral shell (snails), some have a small, flattened shell (semi-slugs) and some species

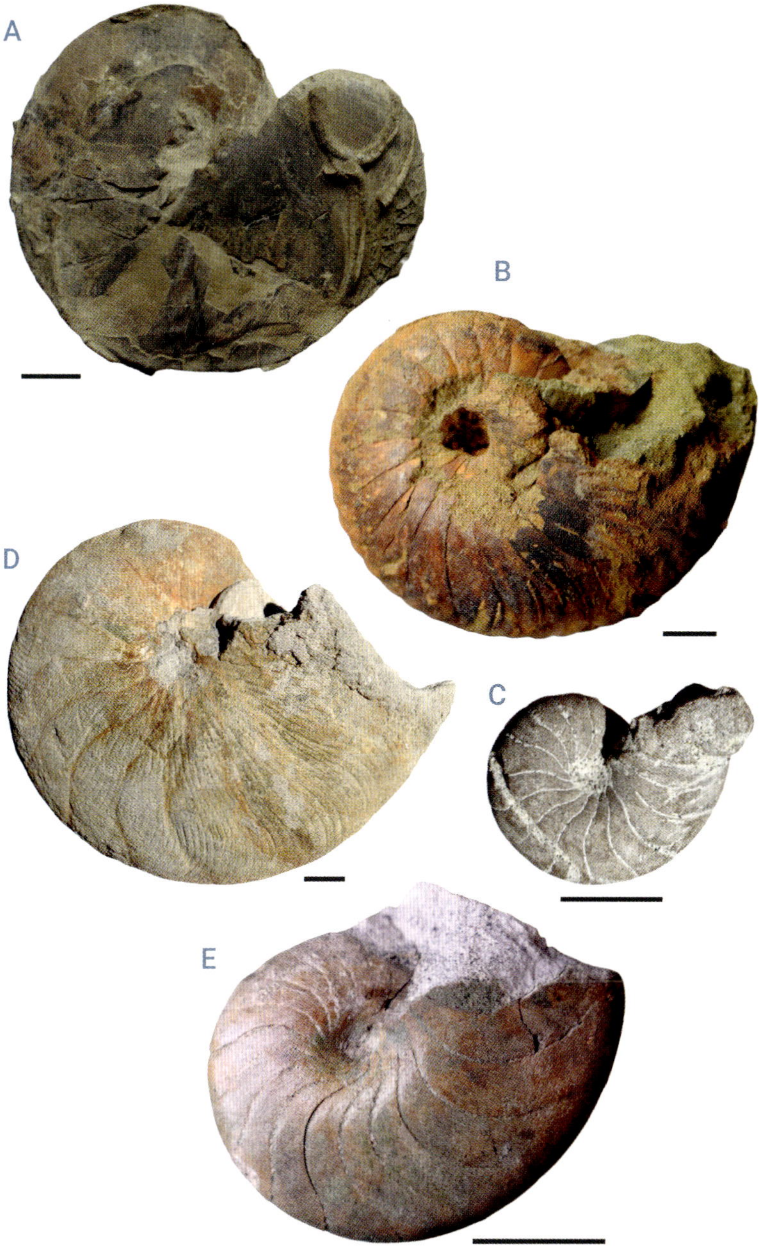

Figure 7.57

Cretaceous nautili.

A, *Heminautilus saxbyi*. Upper Lobster Bed Member, Atherfield Clay Formation, Chale Bay. B, *Cymatoceras radiatum*. Group VI Lower Crinoceras Bed, Ferruginous Sands Formation, Whale Chine. C, *Eutrephoceras expansum*. Glauconitic Marl Member, West Melbury Marly Chalk Formation, Chalk, St. Lawrence. D, *Cymatoceras elegans*. Glauconitic Marl Member, West Melbury Marly Chalk Formation, Chalk. Woody Bay, Ventnor. E, *Psuedocenoceras fittoni*. Glauconitic Marl Member, West Melbury Marly Chalk Formation, Chalk. Rocken End, Niton. A, B, collection of Glenn Martin. C, D, collection of Theo Vickers. E, collection of Jack Wonfor. Scale equals 10 mm.

have lost the shell entirely (slugs). As the shells are the only hard part of the animal (and some opercula), they have more chance of being preserved in the fossil record. Figured here are the common gastropod species you can find in the marine Cretaceous rocks of the island (*Fig. 7.66*).

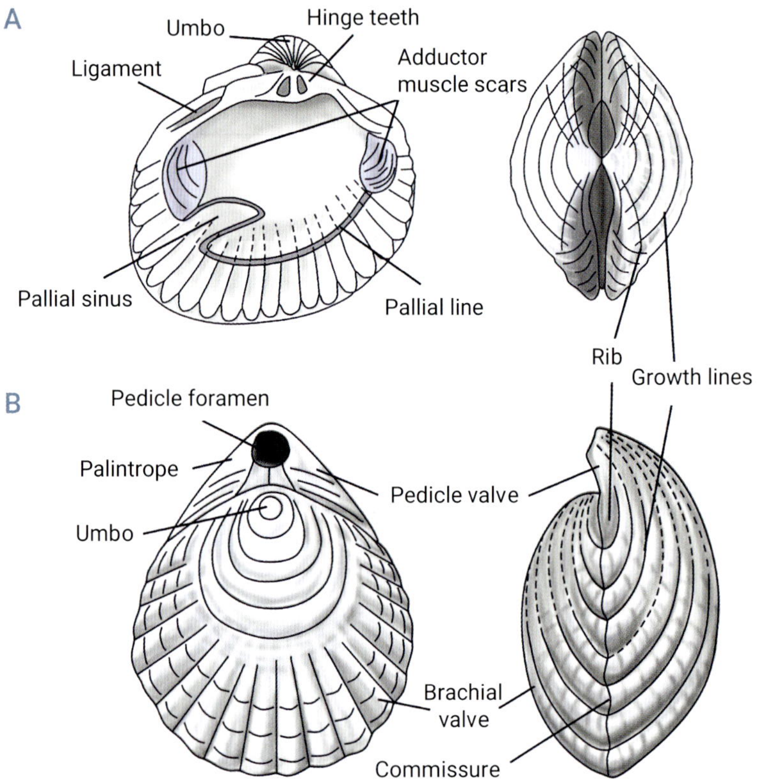

Figure 7.58

Diagram showing A, bivalve and B, brachiopod morphology.

Brachiopods

Brachiopods, despite resembling bivalves, are an entirely different animal, and are only found in marine environments. They consist of two valves (shells), which cover the top and bottom of the animal. Unlike bivalves, where the valves cover the sides of the animal. The valves are unequal in size, with the brachial (top) valve usually being smaller than that pedicle (bottom) valve. The pedicle valve is usually cup shaped, and the brachial valve acts as a lid (*Fig. 7.58*).

There are two main groups of brachiopod, articulate and inarticulate. Articulate brachiopods have a tooth and socket hinge locking the pedicle and brachial valves. Contrastingly, inarticulate brachiopods lack the tooth and socket hinge, and their valves are held together by muscles.

During the Palaeozoic, brachiopods were among the most abundant marine filter feeders, occupying many marine niches.

Figure 7.59

Bivalves from the Perna Bed Member, Lower Greensand Group.

A, *Eurotrigonia caudata*. B, C, *Astarte obovate*. D, E, *Sphaera corrugata*. F, *Modiola aequalis*. G, *Mulletia mulleti*. H, *Trigonia carinata*. I, *Panopea gurgitis*. J, *Notoscabrotrigonia etheridgei*. A – J, collection of Jack Wonfor. Scale equals 10 mm.

However, the Permian – Triassic extinction event saw 66.7% of brachiopods become extinct. Brachiopods took another hit during the end Cretaceous mass extinction event, with 70% extinction of known species in places disappearing. Today, brachiopods are uncommon and low in diversity, with only ~400 species, compared to the ~30,000 known in the fossil record.

In the Cretaceous rocks on the island, it's only possible to find *Lingula*, rhynchonellid and terebratulid brachiopods, all other brachiopod orders became extinct during the Palaeozoic (Fig. 7.67).

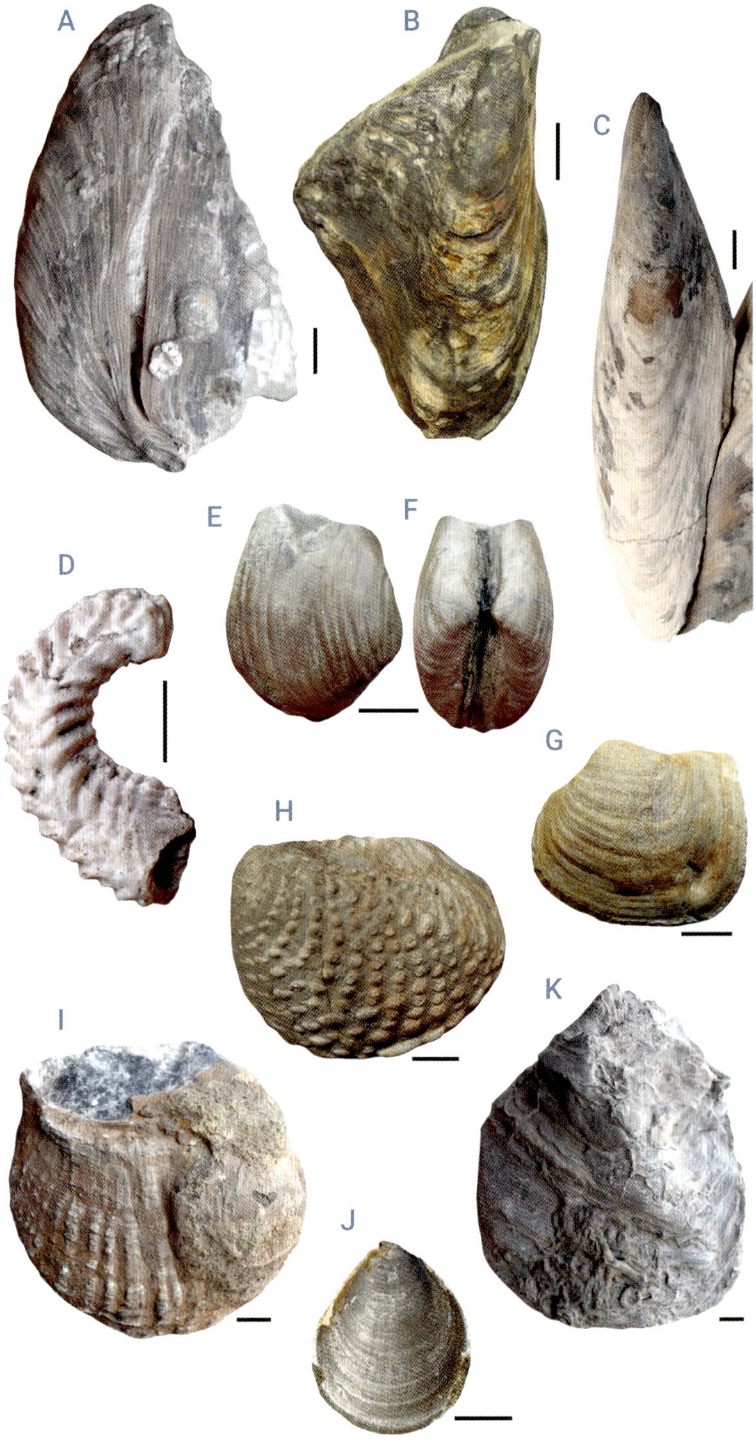

Figure 7.60

Lower Greensand Group bivalves.

A, B, *Gervillaria alaeformis*.
C, *Gervillaria subanceolata*.
D, *Rastellum carinatum*.
E, F, *Panopea gurgitis*.
G, *Panopea mandibula*.
H, *Quadratotrigonia nodosa*.
I, *Prohinnites favrinus*.
J, *Lissochlamys exotica*.
K, *Exogyra sinuata*.
B, H, collection of Glenn Martin. A, C- G, I – K, collection of Jack Wonfor. Scale equals 10 mm.

Figure 7.61

Lower Greensand Group, Upper Greensand and Chalk bivalves. A, *'Cyprina'* sowerbyi. B, Panopea gurgitis. C, *Camptonectes* cottaldinus. D, *Rastellum* sp. E, *Syncyclonema orbicularis*. F, *Chlamys elongatus*. G, *Neithea quinquecostata*. H, *Rastellum* sp. I, L, *Cucullaea obesa*. J, *Neithea quinquecostata*. K, *Neithea quinquecostata*. M, *Ceratostreon tuberculifera*. A, D, F – L, collection of Jack Wonfor. C, Glenn Martin. E, collection of Scott Cade. M, collection of Theo Vickers. Scale equals 10 mm.

Figure 7.62

Shell nodule from the Ferruginous Sands at Shanklin. A, beach rolled nodule. B, water worn nodule in a boulder on the foreshore, showing the cross section of a large bivalve. C, inside of a broken nodule full of *Thetirona* internal and external moulds. D, internal moulds of *Thetirona minor*. E, external mould of *?Cucullea* sp bivalve. Scale equals 50 mm.

Figure 7.63

Bivalves in flint. A, *Pinna* sp. B, *Lima* sp. C, *Neithea sexcostata*. D, internal mould of *Inoceramus* sp. E, broken fragments of ?*Platyceramus*. F, fragment of inoceramid *Cladoceramus* with the boring sponge *Entobia* sp. on the surface. G, external mould of *Inoceramus* sp. Scale equals 10 mm.

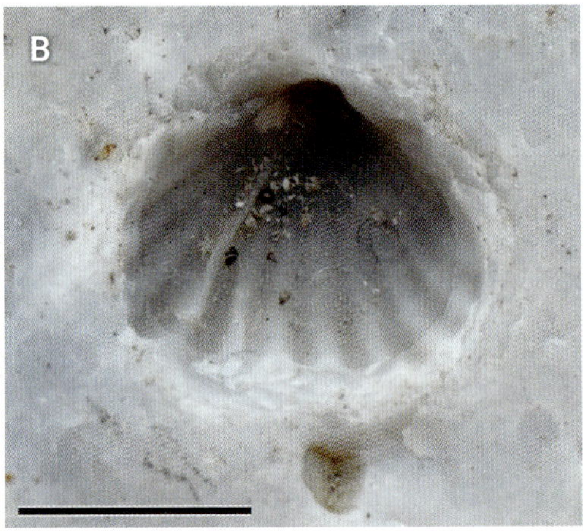

Figure 7.64

Bivalves and brachiopods in flint. A, unknown limacean bivalve. B, *Entolium* sp. C, ?*Microchlamys* scallop. D, internal mould of a terebrachulid brachiopod. E, *Gibbithyris* sp. brachiopod. Scale equals 10 mm.

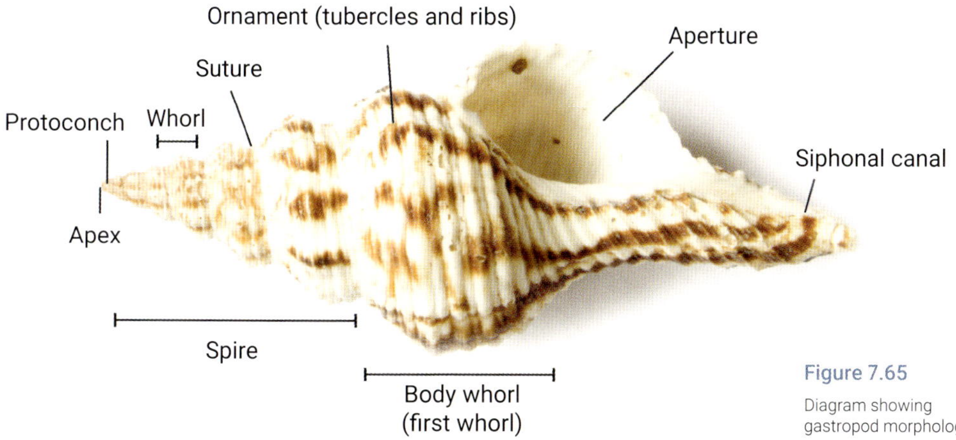

Figure 7.65

Diagram showing gastropod morphology.

Figure 7.66

Cretaceous gastropods. A, *Perissoptera* sp. Ferruginous Sands Formation, Shanklin. B, *Bathrotomaria* sp. Glauconitic Marl Member, West Melbury Marly Chalk Formation, Chalk. C, *Perissoptera glabra*. Crackers member, Atherfield Clay Formation, Chale Bay. D, ?*Torquesia* sp. Crackers member, Atherfield Clay Formation, Chale Bay. E, ?*Pleurotomaria* sp. Glauconitic Marl Member, West Melbury Marly Chalk Formation, Chalk. F, ?*Pleurotomaria vectensis*. Glauconitic Marl Member, West Melbury Marly Chalk Formation, Chalk. G, *Conotomaria gigantea*. Ferruginous Sands Formation, Chale Bay. H, *Perissoptera glabra*. Crackers member, Atherfield Clay Formation, Chale Bay. I, *Pseudomelania* sp. Crackers member, Atherfield Clay Formation, Chale Bay. J, *Gymnentome pizcuetana*. Crackers member, Atherfield Clay Formation, Chale Bay. K, *Conotomaria* cf. *chardstockensis*. Glauconitic Marl Member, West Melbury Marly Chalk Formation, Chalk. B, F, G, collection of Jack Wonfor. C, D, H, I, J, collection of Glenn Martin. E, K, collection of Theo Vickers. Scale equals 10 mm.

Figure 7.67

Cretaceous brachiopods.

A, *Sellithyris sella* 'nest'. Group IV, Ferruginous Sands Formation, Chale Bay. B, *Sellithyris sella*. Group IV, Ferruginous Sands Formation, Chale Bay. C, D, *Moutonithyris dutempleana*. Glauconitic marl member, West Melbury Marly Chalk Formation, Grey Chalk sub-group. E, *Cyclothyris latissima*. Ferruginous Sands Formation, Shanklin. F, *Terebratulina etheridgei*. Glauconitic marl member, West Melbury Marly Chalk Formation, Grey Chalk sub-group. A, B, D, E, F, collection of Jack Wonfor. C, collection of Theo Vickers. Scale equals 10 mm.

Figure 7.68

Cretaceous sponges.

A, *Exanthesis labrosus*, Glauconitic marl member, West Melbury Marly Chalk Formation, Grey Chalk sub-group. B,?*Temnospongia*, Glauconitic marl member, West Melbury Marly Chalk Formation, Grey Chalk sub-group. C, exposure of Lower West Melbury Marly Chalk Formation at Yaverland with D, large *Exanthesis* sp. sponges. A,B, collection of Theo Vickers. Scale equals 10 mm.

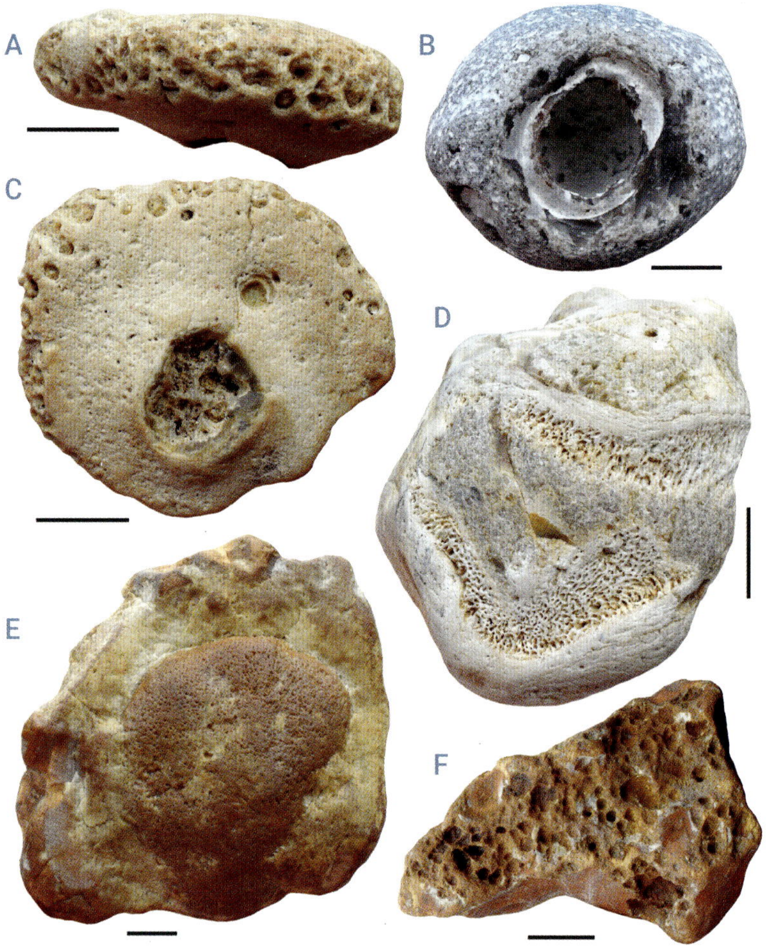

Figure 7.69

Cretaceous sponges.

A, side view and C, ventral view of a possible ?*Ventriculites,* showing the internal structure. B, possible ?*Plinthosella*. D, cross section of a worn sponge in flint. E, internal structure of a sponge showing dense pitting. F, sponge impression. B, private collection of Huxley Hunt. Scale equals 10 mm.

Sponges

Sponges, despite a plant-like appearance, are a simple animal within the phylum Porifera, the name meaning 'pore bearer'. Sponges possess many pores and narrow canals, which are used to funnel water for filter feeding. Sponges are sessile animals, attaching themselves to the seafloor by a root or holdfast-type structure. Some sponges even bore into bivalves, destroying the shell. The oldest fossil sponge dates to the Precambrian, some 600 million years ago.

There are 3 types of sponges: calcareous sponges, glass sponges and demosponges.

Calcareous sponges have spicules made of calcium carbonate and are mainly found in shallow tropical waters worldwide.

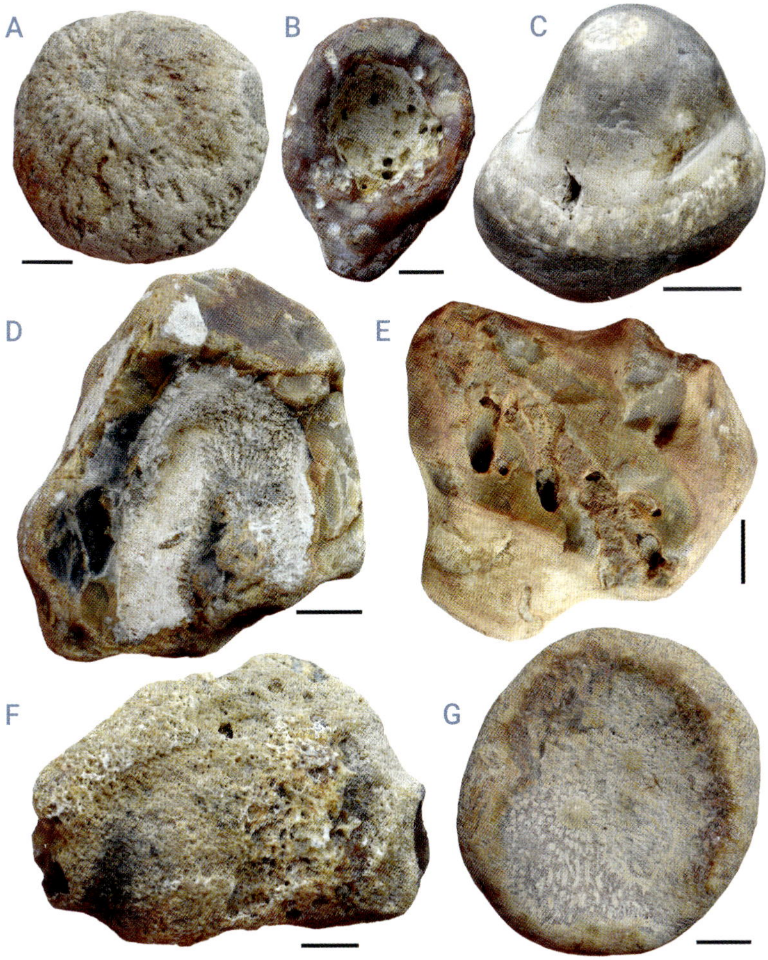

Figure 7.70

Cretaceous sponges.

A, internal structure of a sponge, from the Glauconitic marl member, West Melbury Marly Chalk Formation, Grey Chalk sub-group. B, broken flint sponge showing the internal structure. C, a possible ?*Ventriculites* within a flint pebble. D, impression of the internal structure of a sponge. E, impression of a branching sponge. F, internal structure of a sponge. G, worn surface of a sponge showing internal branching structures. Scale equals 10 mm.

Demosponges or coralline sponges have massive skeletons made of calcium carbonate and are the most diverse class of sponge. Their dense skeletons produce concentric layers, making them look like corals, hence the common name coralline sponges.

Glass sponges or hexactinellid sponges have skeletons made of 4-6 pointed glass-like siliceous spicules (a type of skeleton). These are very delicate cup like animals, mostly living in water depths of 450 – 900 metres. The fossil record of glass sponges is very good, as the individual siliceous spicules preserve well. Some researchers believe these are the longest-living animals on earth, reaching ages of 15,000 years.

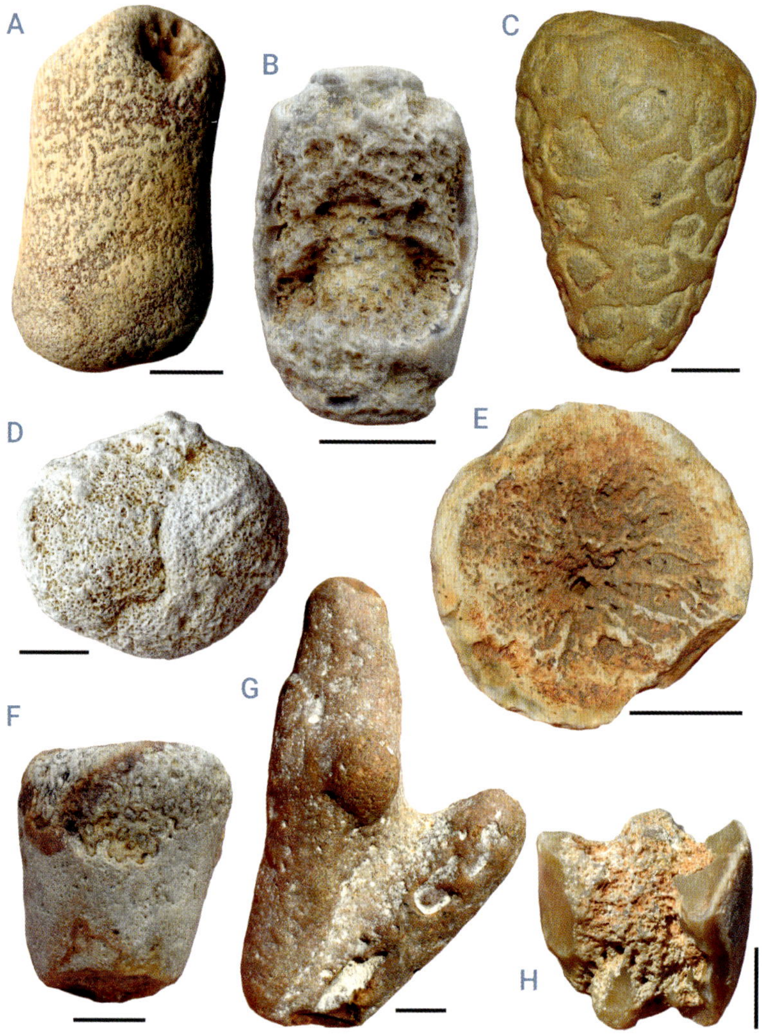

Figure 7.71

Cretaceous sponges

A, internal structure of a sponge. B, possible ?*Sporadoscinia* showing internal structure. C, flint sponge. D, Internal structure of a sponge. E, branching structure inside a flint sponge. F, possible ?*Siphonia*. G, partial ?*Doryderma* showing branching. H, broken sponge showing internal structure. Scale equals 10 mm.

There are also types of sponge, such as *Cliona*, that bore into hard ground or shells that are made of calcium carbonate. They leave shells and rocks covered in holes, with a branching network of canals. These traces are called *Entobia* and are found throughout the fossil record from the Devonian (~400 million years ago) to today.

Sponges are a very common find on the island, with the most abundant being those preserved within flint stones derived from the chalk. They take on a variety of different shapes, from mushrooms, balls, cones, complex branching structures and little round buttons (amongst many others). Sponges can also be found within the chalk,

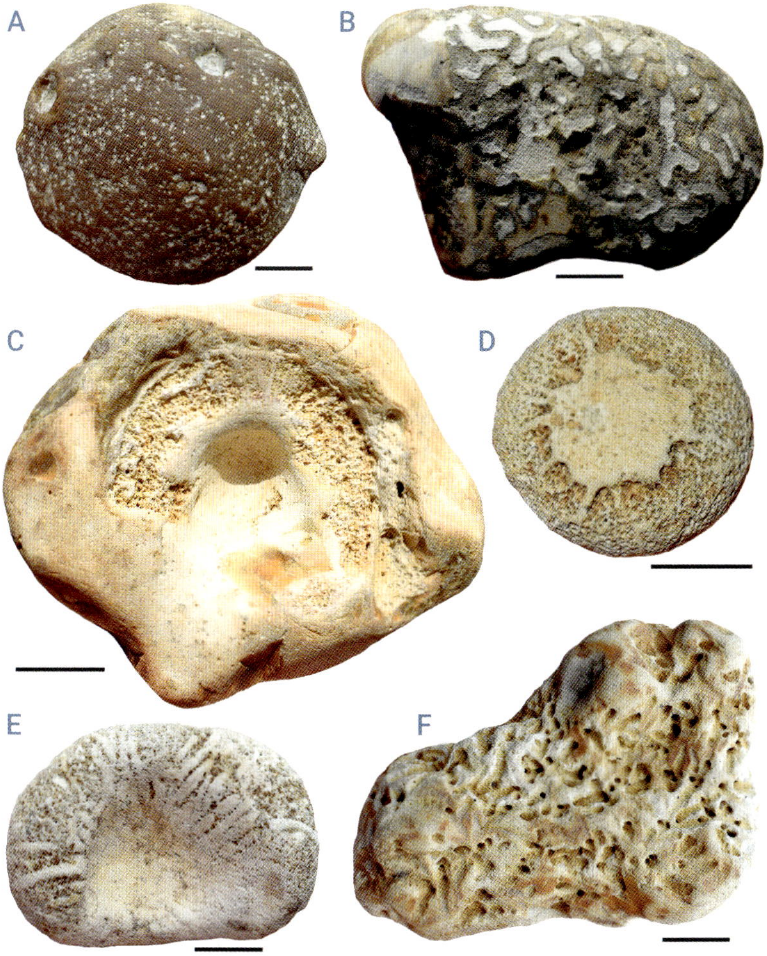

Figure 7.72

Cretaceous sponges.

A, ?*Porosphaera*. B, possible ?*Exanthesis*. C, internal structure of a flint sponge. D, Internal mould of a possible ?*Siphonia*. E, internal structure of a possible ?*Siphonia*. F, casts of a small branching sponge in flint. Scale equals 10 mm.

sometimes as 3D fossils but often only seen due to iron oxide (rust), accumulating within the sponge walls. The Glauconitic Marl also yields lots of sponge fragments (*Figs. 7.68–7.72*).

Echinodermata

Echinoids or sea urchins are a group of echinoderms and thus are related to starfish, brittle stars, sea cucumbers and crinoids. A lot of what we know about extinct echinoids is based on research of living species, with many echinoid groups remaining almost unchanged for millions of years. They are made from a calcite test (shell), consisting of hundreds of small plates fused. The plates are often separated

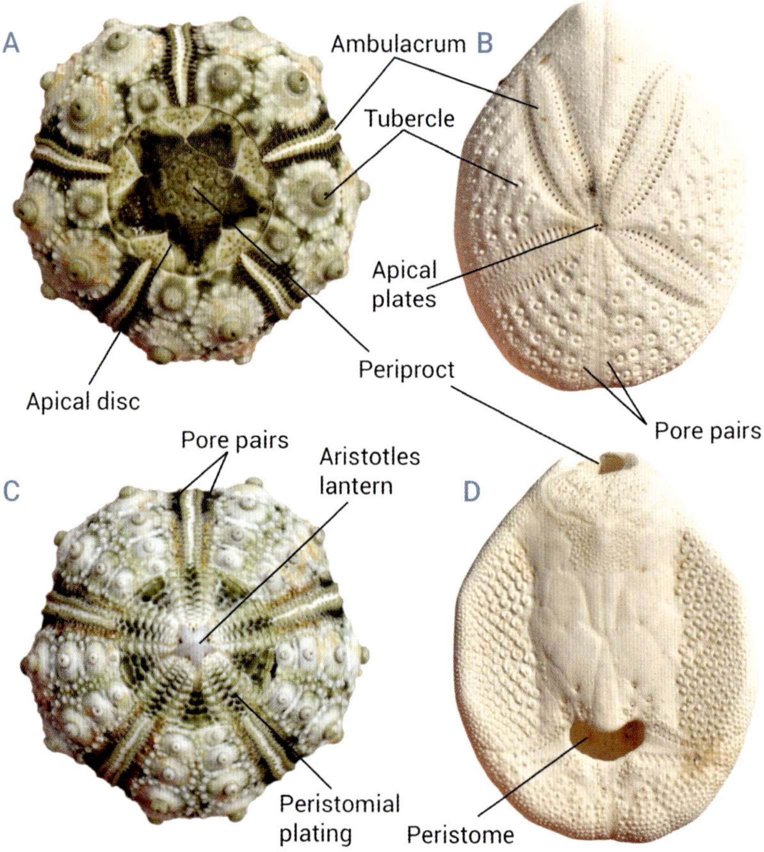

Figure 7.73

Echinoid morphology.

Regular echinoid in A, dorsal and C, ventral views. Irregular echinoid in B, dorsal and D, ventral views.

by narrow bands of pores arranged in pairs. They also have spines attached to the test by soft tissues which are used for protection and locomotion. Tube feet protrude through the pores, resembling suckered tentacle-like appendages and are used to grip onto rocks, locomotion, feeding and respiration (echinoids lack lungs or gills) (*Fig. 7.73*).

The symmetry and star-like pattern on the top of the test have made echinoids curiosities for thousands of years. Fossil echinoids have been found included in Palaeolithic flint hand axes, even Neolithic and Bronze Age burials, and Roman buildings. They have often been referred to as Fairy loaves, after their similarity to traditional loaves of bread, and have been kept as lucky charms or used to ward off evil spirits. They have even been used as weights in ancient markets.

Figure 7.74

Cretaceous sponges.

Echinoids. A, internal mould of *Pseudholaster* sp. from the Upper Greensand Formation. B, internal mould of *Pseudholaster* sp., Ferruginous Sands, Lower Greensand Group. C, *Epiaster* sp. with calcite test preserved Glauconitic Marl Member, West Melbury Marly Chalk Formation, Chalk. D, *Allomma* sp., Glauconitic Marl Member, West Melbury Marly Chalk Formation, Chalk. E, L, spines of *Hirudocidaris hirudo*, White Chalk Group. F, G, *Hyposalenia lardyi*, Lower Grypaea Bed, Ferruginous Sand Formation, Atherfield. H, saleniid regular echinoid. I, flint internal mould of a phymosomatid echinoid. J, *Hirudocidaris* sp., Glauconitic Marl Member, West Melbury Marly Chalk Formation, Chalk. K, *Catopygus* sp. Glauconitic Marl Member, West Melbury Marly Chalk Formation, Chalk.A, B, F, H, collection of Jack Wonfor. C, collection of Dinosaur Expeditions C.I.C. E, collection of Scott Cade. E, L, photo courtesy of Andy Gale. Scale equals 10 mm.

Echinoids are separated into two groups, regular and irregular.

Regular echinoids

Regular echinoids first appeared in the Middle Ordovician (465 million years ago) and saw a significant evolutionary diversification in the Early Jurassic. Many of the Mesozoic groups survived several mass extinctions and still inhabit the seas today. Regular echinoids have a 5-fold symmetry and often display a circular outline when seen from above or below. Cidarids are the most primitive group of echinoids and are the most iconic, with large tubercles and a variety of different spine morphologies.

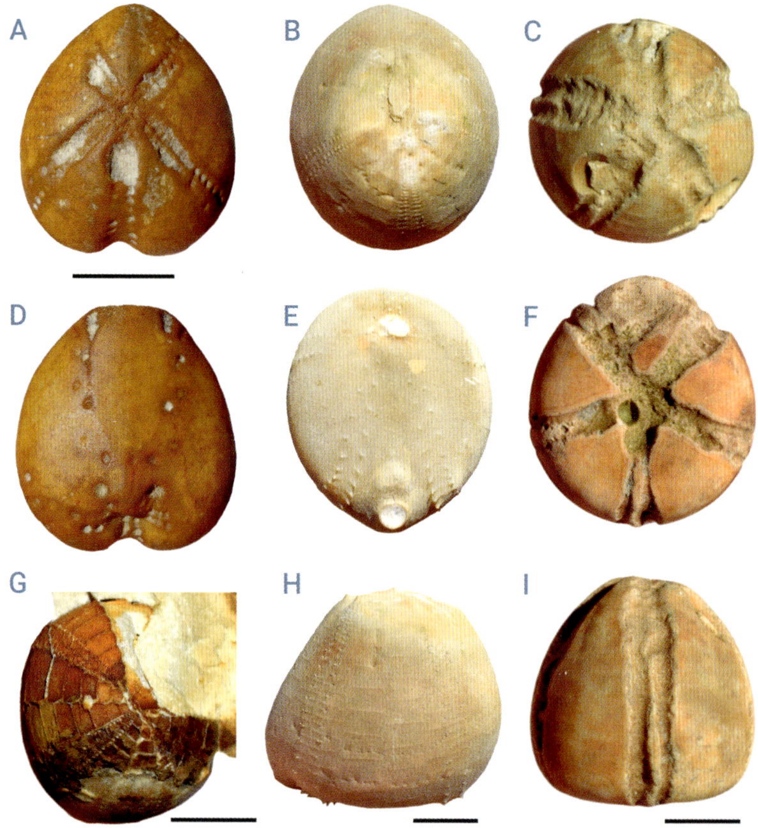

Figure 7.75

Irregular echinoids in flint. A, D, G, *Micraster* sp. B, E, H, *Echinocorys* sp. C, F, I, *Conulus* sp. C, F, I collection of Bournemouth Natural History Society. Scale equals 10 mm.

Most regular echinoids are covered in large spines, attached to the body of the echinoid by ligaments and fine muscles attached to the base of a rounded boss on which the spines sit. Due to only being attached by soft tissues, the spines usually fall off the animal quickly after death. So fossilised remains of echinoids with all the spines attached are a relative rarity and only happen with rapid burial conditions. The spines are used for protection from predation and from rough seas. The animal can move its spines, so can be directed at what they perceive to be a threat. The spines are highly variable in both shape and size (especially length), depending on where the echinoid lived and the threats/pressures present. Cidarids show the most diversity in spine shape, ranging from balloon-like spines, interlocking umbrella-like spines, and very ornate long spines. Some spines also have poison sacs in the tips, an extra deterrent for any potential predator.

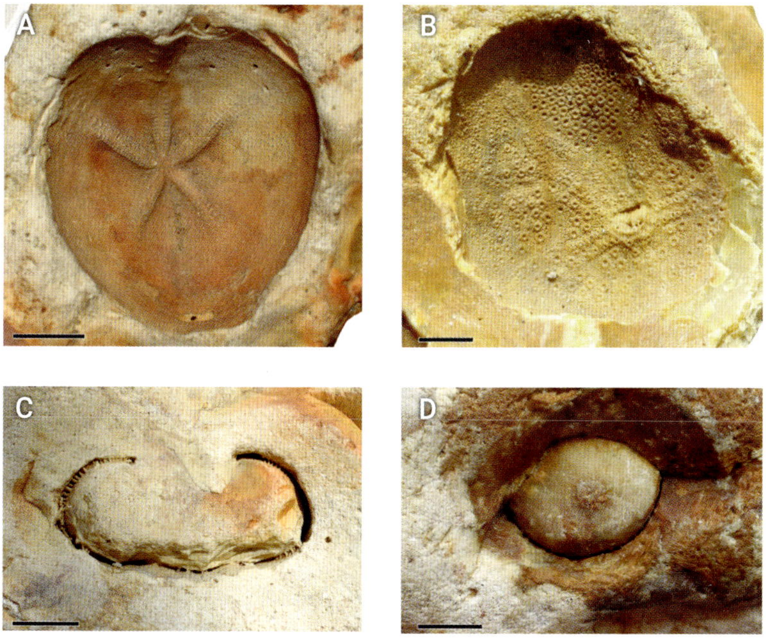

Figure 7.76

Echinoids in flint. A, aboral internal mould. B, oral external mould. C, cross-section of regular echinoid in flint showing flint infilling the pore pairs. D, small echinoid within a flint block. Scale equals 10 mm.

Regular echinoids may be either herbivorous or carnivorous. They possess an Aristotle's lantern (a complex 5-part jaw apparatus) to scrape algae off surfaces whilst grazing and breaking apart food items.

Regular echinoids are usually found as impressions within flints, with a unique pattern composed of rings (resembling squid or octopus suckers). Complete regular echinoids resemble ring doughnuts. Isolated spine impressions are also common finds within flints (*Figs. 7.74, 7.77, 7.78*).

Irregular echinoids

Irregular echinoids, commonly known as sea potatoes, evolved from regular echinoids in the Jurassic, and possess a bilateral symmetry over printed on the 5-fold symmetry of regular echinoids. Typically, irregular echinoids are more common as fossils compared to the regular echinoids, probably because they lived within the sediment (infaunal). They are often found as the internal moulds in flint (*Figs. 7.74-7.76*) Living within the sediment, irregular echinoids are deposit feeders (eating tiny food particles contained within the

Figure 7.77

Regular echinoids in flint. A, impression of external mould of the test of *Temnocidaris* sp. B, broken cidarid echinoid with associated spines. C, impression of external mould of cidarid. D, impression of external mould of *Gauthieria* sp. E, isolated spine of *Temnocidaris* sp. F, partially articulated test and spines of *Tylocidaris clavigera* in flint. F, Dinosaur Expeditions C.I.C collection. Scale equals 10 mm.

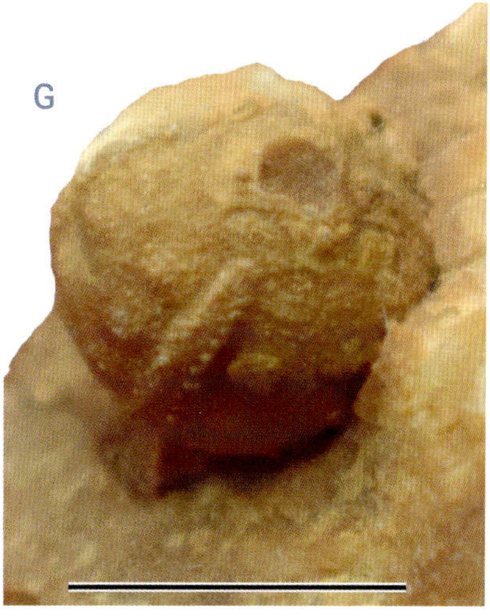

Figure 7.78

Regular echinoids in flint. A, ventral (bottom) test impression and B, internal mould of *Phymosoma* sp. C, impression of dorsal (top) surface of *Salenia geometrica*. D, exposed internal mould of regular echinoid. E, F, impressions of *Tylocidaris* sp. spines. G, internal mould of *Salenia* sp. A, B, Bournemouth Natural History Society collections. C, G, collection of Rick Lilley and Charlotte Willis. F, collection of Scott Cade. Scale equals 10 mm.

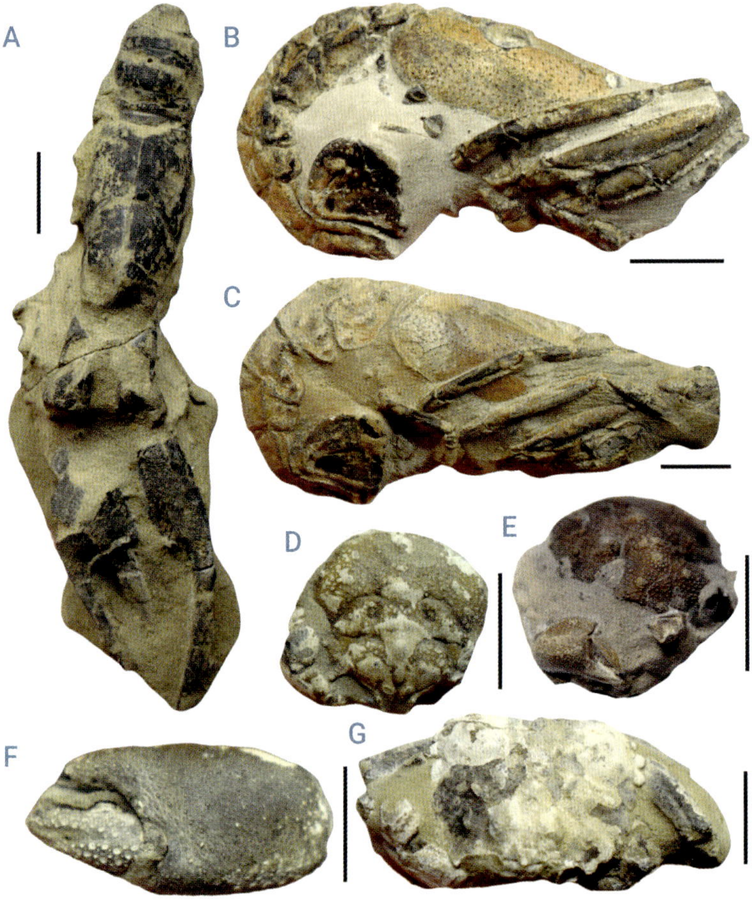

Figure 7.79

Lower Greensand crustaceans.

A, *Hoploparia longimana*. B, *Atherfieldastacus magnus*. C, *Atherfieldastacus magnus*. D, *Mithracites vectensis*. E, *Mithracites vectensis*.F, isolated crab claw. G, currently unidentified species of crab. All specimens are from the Lower lobster bed, Atherfield Clay Formation, Chale Bay. A-D, F, G, collection of Glenn Martin. E, collection of Jack Wonfor.

sediment). Consequently, they churned up the chalk, destroying primary sedimentary structures. As a result of their infaunal lifestyle, irregular echinoids spines are greatly reduced and thinned to aid in their locomotion whilst burrowing and generate currents within the burrow rather than for defence purposes. The spines are also directed posteriorly to streamline their passage through the sediment.

The irregular echinoids are hard to miss, having a star composed of pairs of dots on the top and are either domed or heart shaped.

Crustaceans

Crustacean fossils are common on the Isle of Wight. The Lower Lobster Bed at Atherfield is famous for the high abundance of well-

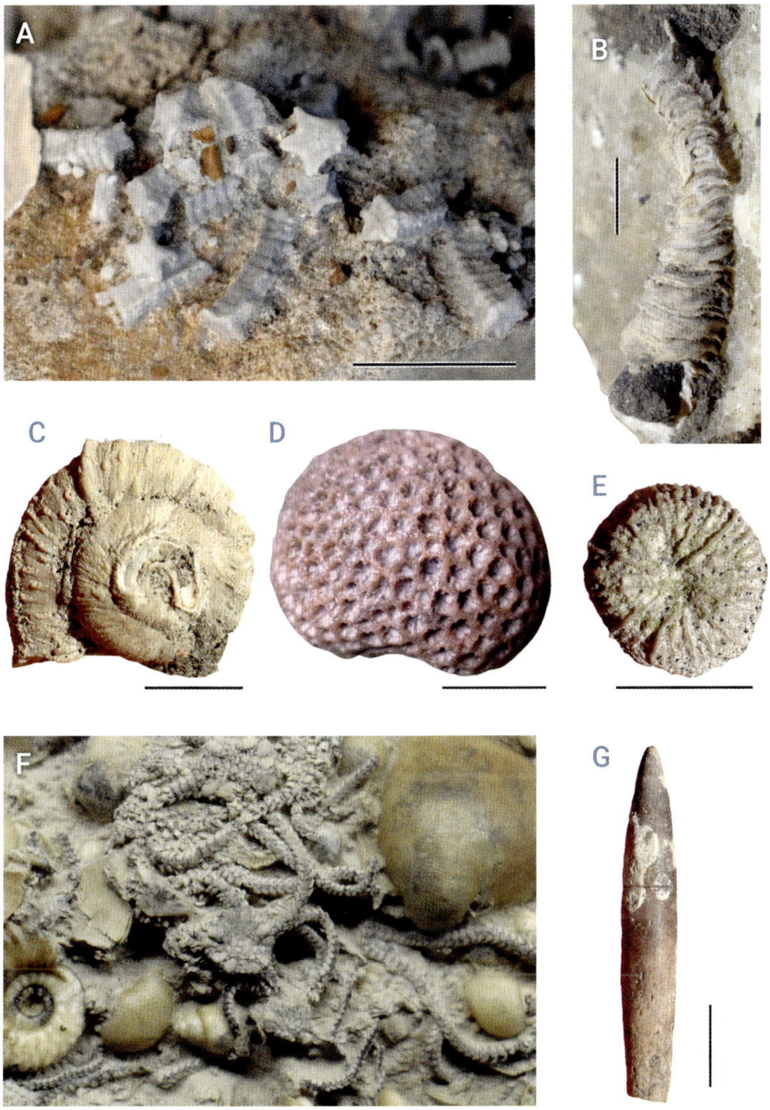

Figure 7.80

Other marine Cretaceous invertebrates.

A, isocrinid ossicles (crinoid stem segments). Lower Greensand Group. B, serpulid tube, Ferruginous Sands Formation, Chale Bay. C, serpulid worm, *Rotularia* (*Praerotularia*) *umbonata*. Glauconitic marl member, West Melbury Marly Chalk Formation, Grey Chalk sub-group. D, colonial coral, *Holocystis elegans*, Perna Bed Member, Lower Greensand Group, Atherfield. E, solitary coral, *Micrabacia coronula*. Glauconitic marl member, West Melbury Marly Chalk Formation, Grey Chalk sub-group. F, undescribed ophiuroid (brittle Star). Crackers member, Atherfield Clay Formation, Chale Bay. G, belemnite, Glauconitic marl member, West Melbury Marly Chalk Formation, Grey Chalk sub-group Glauconitic Marl. A, B, D, collection of Jack Wonfor. C, E, G, collection of Theo Vickers. F, collection of Glenn Martin. Scale equals 10 mm.

preserved lobsters, interpreted to be mostly the shed exoskeletons, but contain the original colour pigments. Experiments showed that when heating up the pigments, they turned from blue-grey to red. Which is the same process modern lobster shells undergo when cooked. These lobsters, called *Atherfieldastacus magnus*, are often found complete, with the tail tucked under the abdomen. Lobsters can be found in the Upper Lobster Bed Member but tend to be more fragmentary.

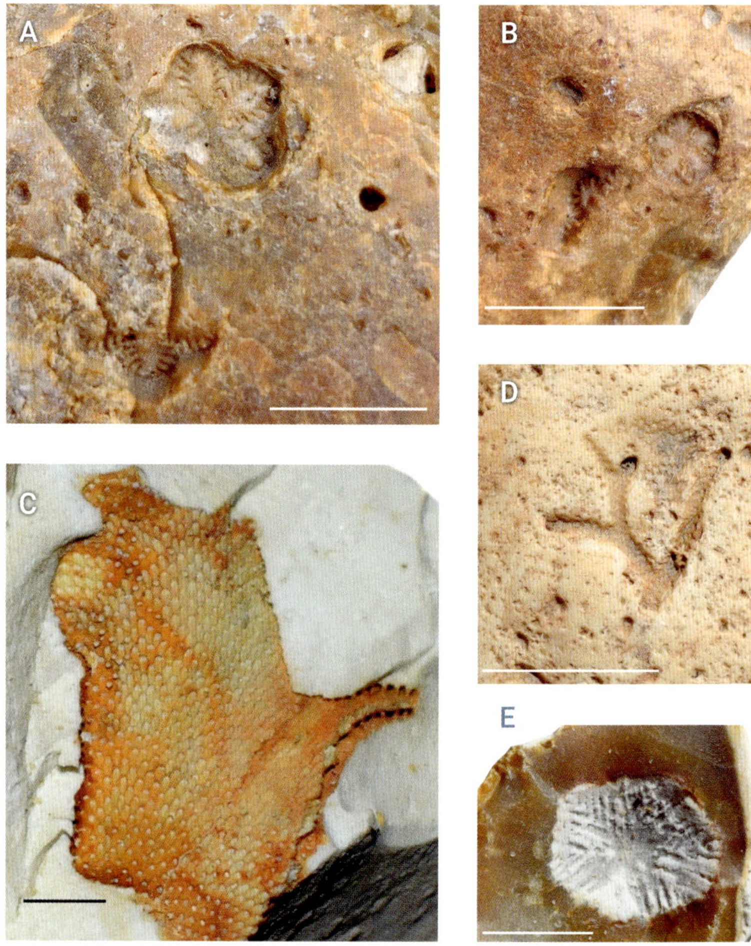

Figure 7.81

Other fossils in flints.

A,B, *Isocrinus* sp. crinoid ossicles. C, encrusting bryozoan. D, branching bryozoan. E, calyx plate from *Marsupites*, a stemless crinoid. Scale equals 10 mm.

Crabs are also found within the Lower Greensand Group but tend to be small, and a rarer find compared to the lobsters (*Fig. 7.79*).

Other invertebrates

Serpulid worms

Serpulid worms are a type of tube-building worm, with the oldest dating to the Permian (290 million years ago). These worms secrete calcite to form hard shells for protection. They also have a special operculum which acts as a door, blocking the entrance to the tube when the worm withdraws. Serpulid worms either encrust onto rocks or shells, cement themselves together to form reefs or live

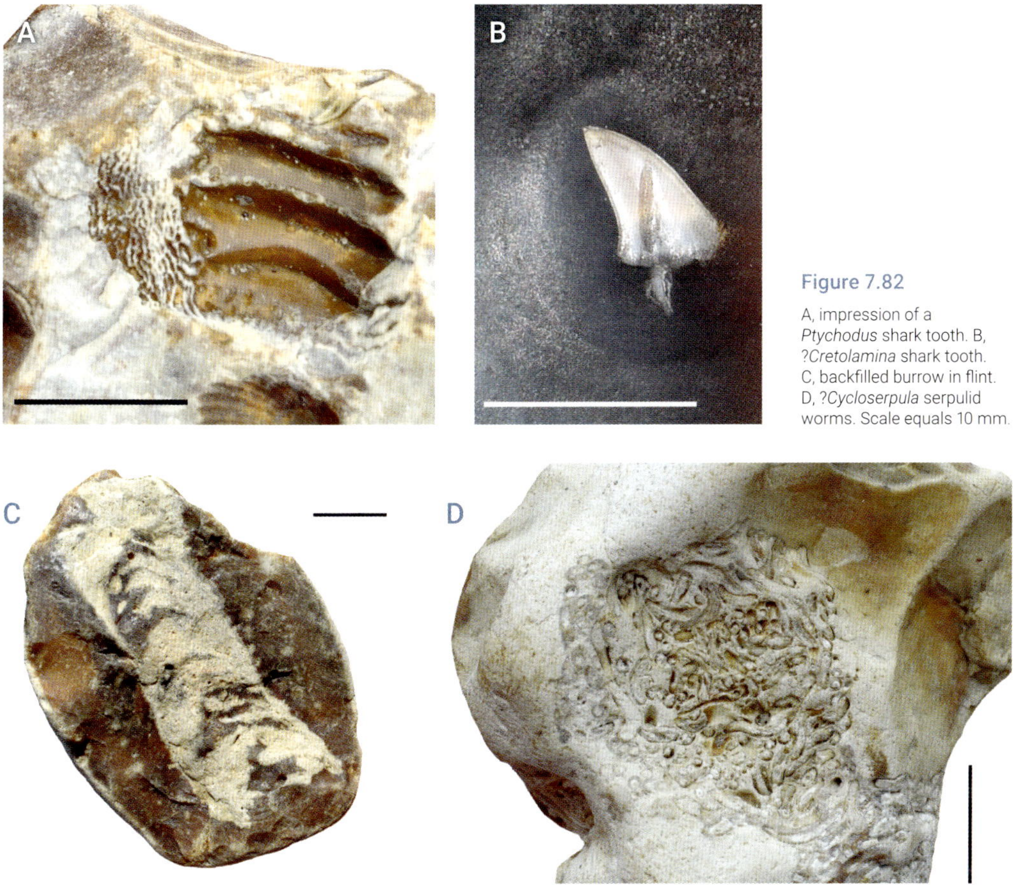

Figure 7.82

A, impression of a *Ptychodus* shark tooth. B, ?*Cretolamina* shark tooth. C, backfilled burrow in flint. D, ?*Cycloserpula* serpulid worms. Scale equals 10 mm.

as isolated individuals. Their tubes may be straight, gently curved, spiral or even form meandering tangled masses.

The spiralled forms are often confused with gastropods (snails); however, a crucial difference is that serpulid worms grow as a tube, creating a double layer between the spirals. Whereas gastropods grow around an axis, creating a column-like structure inside and only one layer of shell (see *Fig. 7.65*).

Serpulid worms are commonly found throughout the Lower Greensand Group and the glauconitic marls, either encrusting onto other fossils or as isolated tubes (*Fig. 7.80, 7.82*).

Crinoids

Crinoids, often called sea lillies, are a type of echinoderm, related to starfish and echinoids. Crinoids attach themselves to the seafloor by a holdfast, similar to that of kelp seaweeds. The main body of the animal consists of a columnal stema, a cup called a calyx and many branching arms with pinnules. The internal skeleton is made from hundreds of calcite plates held together with soft tissues. The plates of the stem can be circular or star-shaped with a central hole. In many places in England, the individual circular plates were termed 'fairy money', due to their resemblance to coins.

The fossil record of crinoids begins around 485 million years ago in the Ordovician. So far over 6000 fossil species have been named and 660 living species have been discovered, making crinoids an incredibly diverse and successful group of animals.

They occur commonly in flints either as small star-shaped plates or partial articulated columns, which can appear as a hollow column full of stacked disks. They also occur as more articulated fossils within cracker nodules at Atherfield (*Fig. 7.80, 7.81*).

Bryozoans

Bryozoans, like corals, are a colonial animal, but are more closely related to brachiopods than to corals. This is because they have a lophophore, which is a specialised feeding organ consisting of a ring of tentacles surrounding the mouth, used for suspension feeding.

Bryozoa form colonies consisting of thousands of zooids (individuals). The zooids secrete calcium carbonate, chitin, or both to form a protective 'box' to live within and are typically only 0.5mm long.

Bryozoans are very diverse organisms with 15,000 fossil species and a fossil record dating back to the Ordovician some 480 million years ago. Their colonies can form seaweed-like structures, others form complex fans, and some encrust onto hard surfaces or even bore into shells.

Due to their hard exoskeleton, bryozoans preserve well and are commonly found with flint stones and encrusting other hard fossils. These fossils are best seen with magnification using a hand lens or magnifying glass (*Fig. 7.81*).

Corals

Corals are a type of marine invertebrate, typically living in colonies of identical polyps, and are within the phylum Cnidaria, making them closely related to jellyfish and sea anemones. The polyp is the living part of the coral, which secretes a calcium carbonate exoskeleton, usually aragonite, to build a shell. Colonies form when many (sometimes millions) of polyps encrust together. Over time they can form huge colonies, producing structures sometimes metres in diameter, with hundreds, if not thousands of colonies forming reefs. The polyps feed by using their tentacles to catch plankton, very similar to anemones, as well as having a symbiotic relationship with photosynthetic dinoflagellates (brown algae), which use sunlight to produce energy.

The fossil record of corals begins in the Cambrian, some ~500 million years ago, with a complex record of radiations and extinctions. Two main groups of corals appear in the Palaeozoic, tabulate and rugose corals, both of which become extinct at the Permian – Triassic boundary. Modern corals, known as Scleractinian corals, appeared in the early Triassic. During the Cretaceous, the world was largely devoid of coral reefs. Instead, reefs were dominated by rudist bivalves. However, there were still many isolated corals known, with well over 1000 species described, mainly from mid-Cretaceous and younger strata (correlating with a global sea level rise). It was only after the extinction of rudist bivalves in the very late Cretaceous that coral reefs started forming again.

Most corals found on the island come from the Cretaceous Perna Bed, outcropping at Atherfield, Yaverland and Compton (*Fig. 7.80*). These are normally the small compound coral *Holocystis* which are small ball-shaped corals. There are also rare and important specimens of *Acropora*, commonly known as the staghorn coral, at Whitecliff Bay. Today, *Acropora* is the most abundant coral in the world and is under threat due to bleaching and environmental destruction.

Fish

Bony fish

Despite being marine deposits, fish fossils are rare. However, their remains do occur in the form of isolated teeth and vertebrae. Only rarely does an articulated skull of a complete fish occur. In the Lower Greensand, large, pointed teeth occur and are referred to the large swordfish-like fish, *Protosphyraena*. However, these remains likely belong to an as of-yet undescribed taxa, as *Protosphyraena* is an Upper Cretaceous fish (*Fig. 7.83*).

Cartilaginous fish

Shark teeth are commonly found throughout the Lower Greensand Group to the Upper Greensand Formation, ranging from 5mm to several centimetres long (*Fig. 7.82, 7.83*).

Reptiles

Reptile remains are surprisingly rare within the marine Cretaceous rocks on the island, despite there being likely many inhabiting the seas and skies when these strata were being deposited. However, there have been remains of turtles, marine reptiles, pterosaurs, and even terrestrial dinosaurs found. To identify bone, *see the page 146*. Due to the rarity of reptiles in the marine rocks, it's always worth reporting your finds to the local museum.

Several dinosaurs, including *Mantellisaurus atherfieldensis* and *Vectaerovenator*, as well as other unnamed ornithopod and sauropod material have been reported from the Lower Greensand Group on the island. These dinosaurs would have been swept out to sea, perhaps drifting miles from the land before eventually sinking to the seafloor. Whilst the animal's carcass was afloat at sea, as well as on the seafloor prior to burial on the seafloor, there was likely a large amount of scavenging, resulting in very incomplete remains being preserved.

Pterosaur remains have also been found, often isolated elements, with one named species *Vectidraco daisymorrisae*, being found in the Atherfield Clay Formation at Atherfield Point. This pterosaur was a small animal with an estimated wingspan of just under a metre and known from a partial pelvis and some vertebrae. For more on pterosaurs, *see page 175*.

A variety of marine reptiles, including crocodyliforms, ichthyosaurs, pliosaurs and plesiosaurs have been found within the Lower Greensand Group and chalk formations on the island. These include isolated bones, teeth, and occasionally a more articulated specimen. It has been suggested that some of the marine reptile material found is derived from the Late Jurassic Kimmeridge Clay, which was eroding at the time the Lower Greensand was being deposited (*Fig. 7.83*).

Ichthyosaur vertebrae are easy to identify, being round with an hourglass shaped cross section, very similar to shark vertebrae. Ichthyosaur teeth are conical and have fine ridges running top to bottom of the crowns.

Trace fossils

There are many trace fossils within the marine rocks on the island, indicating that there was a very productive seafloor community. The sandstones of the Lower Greensand are often rich in trace fossils, mostly represented by a diverse assemblage of burrows, from complex backfilled burrows to simple tube burrows. There are 3D pellet lined burrows commonly found in the Chale Clay member, Atherfield Clay, at Atherfield point. These are often referred to as *Ophiomorpha* - crustacean (shrimp, crayfish, lobster or crab) burrows, lined with 'spit balls' or coprolites (*Fig. 7.85*).

Flint often infills burrows within the chalk and can be seen as large complex structures in the flint beds, or as loose pebbles on the beach. These burrows are often referred to as *Thalassinoides*, often associated with decapod crustaceans (crabs, shrimp, crayfish and lobsters) (*Fig. 7.82*).

Paramoudras or potstones are also found in the chalk at Whitecliff Bay. These are hollow centred, doughnut shaped flint nodules. The most impressive paramoudras are seen at West Runton, North Norfolk, with some reaching 2 metres across and 6 metres high! These are often said to be made by sponge reefs, however

Figure 7.83

Vertebrate remains.

A, unidentified fish skull. West Melbury Marly Chalk Formation, Grey Chalk sub-group. B, *Protosphyraena* fish tooth. Lower lobster bed, Atherfield Clay Formation, Chale Bay C, partial shark vertebrae. Glauconitic marl member, West Melbury Marly Chalk Formation, Grey Chalk sub-group. D, marine reptile bone fragment. Glauconitic marl member, West Melbury Marly Chalk Formation, Grey Chalk sub-group. E, *Cretoxyrhina mantelli*. Glauconitic marl member, West Melbury Marly Chalk Formation, Grey Chalk sub-group. F, ?plesiosaur tooth. Lower lobster bed, Atherfield Clay Formation, Chale Bay. G, pliosaur tooth, Lower Greensand Group with H, in ventral view showing pulp cavity. I, *Cretalamna appendiculata*. West Melbury Marly Chalk Formation, Grey Chalk sub-group. J, *Cretalamna appendiculata*. Glauconitic marl member, West Melbury Marly Chalk Formation, Grey Chalk sub-group. K, *Cretoxyrhina mantelli*. Glauconitic marl member, West Melbury Marly Chalk Formation, Grey Chalk sub-group. A, D, E, I, J, collection of Theo Vickers. C, K, collection of Jack Wonfor. F, collection of Huxley Hunt. G,H, collection of Claire Hempstead. A, B, C, D, G, H, I, scale equals 10 mm. E, F, J, K, scale equals 5 mm.

Figure 7.84

Examples of pebbles from the Lower Greensand containing fossils.

A, serpulid worms. B, fragments of bivalve, gastropods, ammonites and other shelly fossils. C, section of crinoid with serpulid worms. D, section of ammonite showing calcite sutures. A, collection of Huxley Hunt. B – D, collection of Scott Cade. Scale equals 10 mm.

this is not the case. There are two possibilities for the formation of paramoudras, either the flint has formed around a deep vertical animal burrow structure, or they are related to gas escaping the sediment, effecting the formation of the surrounding flint (*Fig. 7.85*).

Botany

Despite the marine rocks on the island being deposited miles away from the nearest landmass, plant material is very common within the Sandrock at Shanklin. The fossilised wood found would have been driftwood, washed out to sea by rivers draining from a landmass in the northeast. Evidence of the time spent at sea are the countless borings, referred to the ichnogenus *Teredolites*. These are likely the result of a wood-boring bivalve in the family Xylophagaidae. These wood-boring bivalves would have made a hole in the wood using their shells as a file, then filter feeding as the wood drifts in the currents, until it eventually sinks to the seafloor. Some pieces of fossil wood are completely riddled with traces of these bivalves.

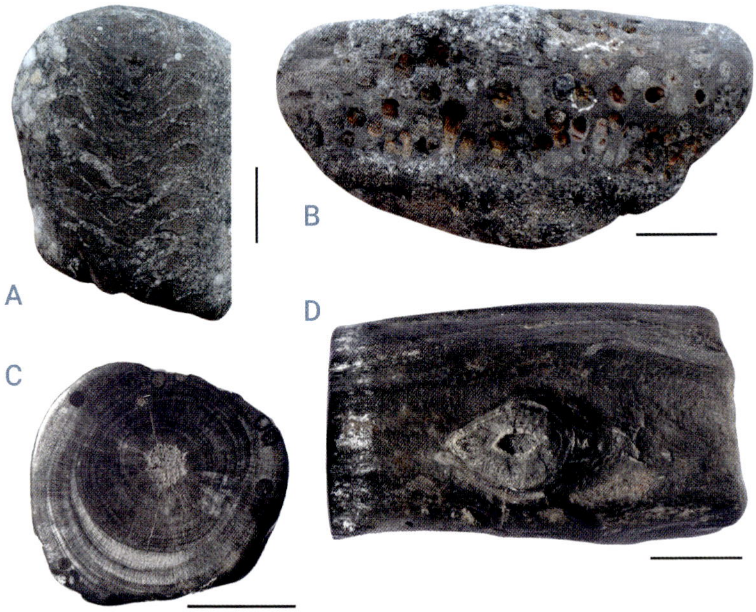

Figure 7.86

A, cross section of a pinecone in sandstone. B, wood with *Teredolites* borings. C, polished section of wood showing internal borings and tree rings. D, section of wood with a branch attachment. B, D, collection of Scott Cade. Scale equals 10 mm.

The fossilised driftwood is very common at Shanklin but can also be found at Yaverland and Compton. The occasional piece of wood, preserved without borings, can be found within the Lower Greensand Group at Atherfield (*Fig. 7.86*).

Silicified wood also occurs occasionally within the Upper Greensand Formation, and incredibly rare pieces of amber occur within the chalk.

7.5. Paleogene

The Paleogene rocks on the island span some ~ 30 million years from 56 to 32 million years ago. They encompass a variety of rock types

Figure 7.87

Nummulites. A, multiple fossils complete in matrix. B, internal structure. Scale equals A, 10 mm, B, 5 mm

Figure 7.88

Eocene bivalves.

A, B, *Venericor planicosta*, Bracklesham Group. C, F, *Artica planata*, London Clay Formation. D, E, *Polymesoda convexa*, Bembridge, Solent Group. G, H, *Bathytormus sulcatus*, Barton Group. I, J, *Polymesoda obovata*, Bembridge, Solent Group. K, L, *Glans sp.*, Barton Group. M, N, *Polymesoda obtusa*, Thorness, Solent Group. O, P, *Cubitostrea ventilabrum*, Solent Group. Q, R, *Glycymeris* sp., Barton Group. S, T, *Polymesoda* sp. Solent Group. U, V, *Cubitostrea plicata*, Barton Group. O, P, collection of Theo Vickers. U, V, collection of Rick Lilley and Charlotte Willis. A - H, S - V, scale equals 10 mm. I - N, Q, R, scale equals 5 mm.

representing many different sedimentary environments. For ease of identification, all the fossils have been put into their respective groups into age order.

Foraminifera

Foraminifera or forams are single-celled organisms, commonly with an external shell or test composed of calcite, which is what is preserved in the fossil record. The tests can be very elaborate with multiple chambers. Most forams require sieving of sediment and viewing under a microscope to be able to see them in detail, only reaching a couple of millimetres in size. However, there is one foram, *Nummulites*, which grows to an incredible size (for forams), ranging from 1.3 cm – 5 cm in diameter (*Fig. 7.87*). Superficially, these can look like worn pennies but once split in half a spiralled chambered structure can be seen internally. The occurrence of

Figure 7.89

Eocene shell beds.

A, section of foreshore shell bed from Thorness, showing colour banded *Polymedsoda convexa*, pyritised internal moulds and gastropod rich matrix. B, iron concretion with *Tarebia acuta* gastropods. C, iron concretion with *Viviparus* gastropods. D, shell bed at Thorness with occasional 1, *Melanopsis carinata*. dominated by 2, *Potamaclis turritissima*. E, Nematura bed at Bouldnor showing *Polymedsoda convexa* with colour banding and the gastropods *Tarebia acuta* and *'Potamides'* sp. Scale equals 10 mm.

Nummulites has puzzled many foram workers, as they usually only occur in rocks laid down in the Tethys Sea to the south in limestones but are incredibly common here in at a more northern latitude in sandstones.

Bivalves

The bivalve faunas of the Paleogene rocks of the island are highly diverse, representing a variety of different saltwater, brackish and

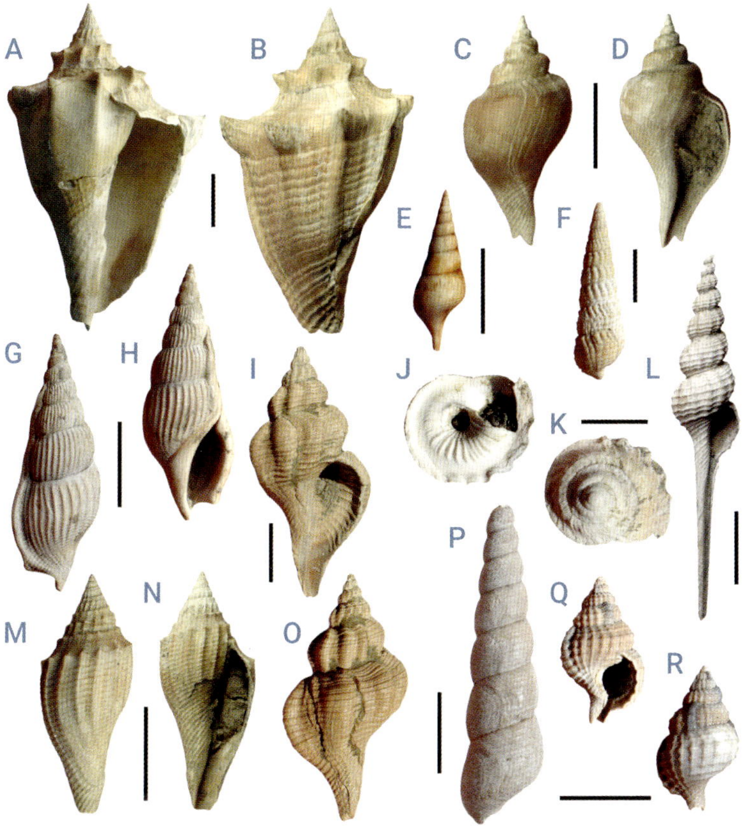

Figure 7.90

Eocene gastropods.

A, B, *Volutospina luctator*, Barton Beds. C, D, *Sycostoma pyrus*, Barton Group. E, *Hippochrenes amplus*, Barton Group. F, *Granulolabium plicatum moniliferum*, Thorness, Solent Group. G, H, *Rimella rimosa*, Barton Group. I, O, *Wrigleya regularis*, Barton Group. J, K, *Xenophora schroeteri*, Barton Group. L, *Fusinus porrectus*, Barton Group. M, N, *Volutospina luctator*, Barton Group. P, *Haustator editus*, Barton Group. Q, R, *Adetula evulsa*, Barton Group. A - D, M, N, collection of Rick Lilley and Charlotte Willis. I, O, collection of Theo Vickers. Scale equals 10 mm.

freshwater environments (*Fig. 7.88*). At some points in time, certain species of bivalves flourished, so much so that they formed shell beds, which can now be seen in several locations along the coastline (*Fig. 7.89*). It's also worth noting the shell beds are also a good source of vertebrate fossils. At some localities on the island, the original colour banding of bivalves is preserved.

Gastropods

Gastropods, commonly known as slugs and snails, are found commonly throughout the Palaeogene strata on the island, ranging from large marine species to ornate freshwater and brackish species. Incredibly, at some localities, specimens with the original colour banding still preserved can be found. The gastropod fauna is highly diverse, with only the most common species figured here (*Figs. 7.90, 7.91*), but a full list of the molluscan faunas can be found on Alan Morton's DMAP fossils website.

Figure 7.91

Eocene gastropods.

A, B, *'Filholia'* sp., Headon Hill. C, D, *Headonia eumphalus*, Headon Hill, Solent Group. E, *Clavilithes pinus*. Barton Group. F, G, internal mould of *Trochopaludina angulosus*, Headon Hill, Solent Group. H, *Lymnaea* sp, Headon Hill, Solent Group. I, J, *Viviparus lentus*, Headon Hill, Solent Group. K, L, *Tarebia acuta*, Thorness, Solent Group. M, *Potamides (Ptychopotamides) vagus*, Headon Hill, Solent Group. N, *Tympanotonos* sp., Thorness, Solent Group. O, *Vicinocerithium concavum*, Headon Hill, Solent Group. P, *Haustator* sp., Bracklesham Group. D, H, I, J, collection of Theo Vickers. M, collection of Rick Lilley and Charlotte Willis. Scale equals 10 mm.

Fish

Bony fish

Fish remains are common from the Paleogene strata on the island, skull bones including jaw fragments and teeth, vertebrae, scales, fin spines all occur frequently. The fish remains are often referred to the genus *Amia* sp., otherwise known as the bowfin, and a gar called *Lepisosteous* is also present. The fish fauna is likely very diverse, however, the fragmentary nature of the fossils doesn't allow for easy identification.

Vertebrae are easy to identify, being circular to oval with a dished surface front and back, and often with prominent ridges around the edges. Depending on the position of the vertebrae

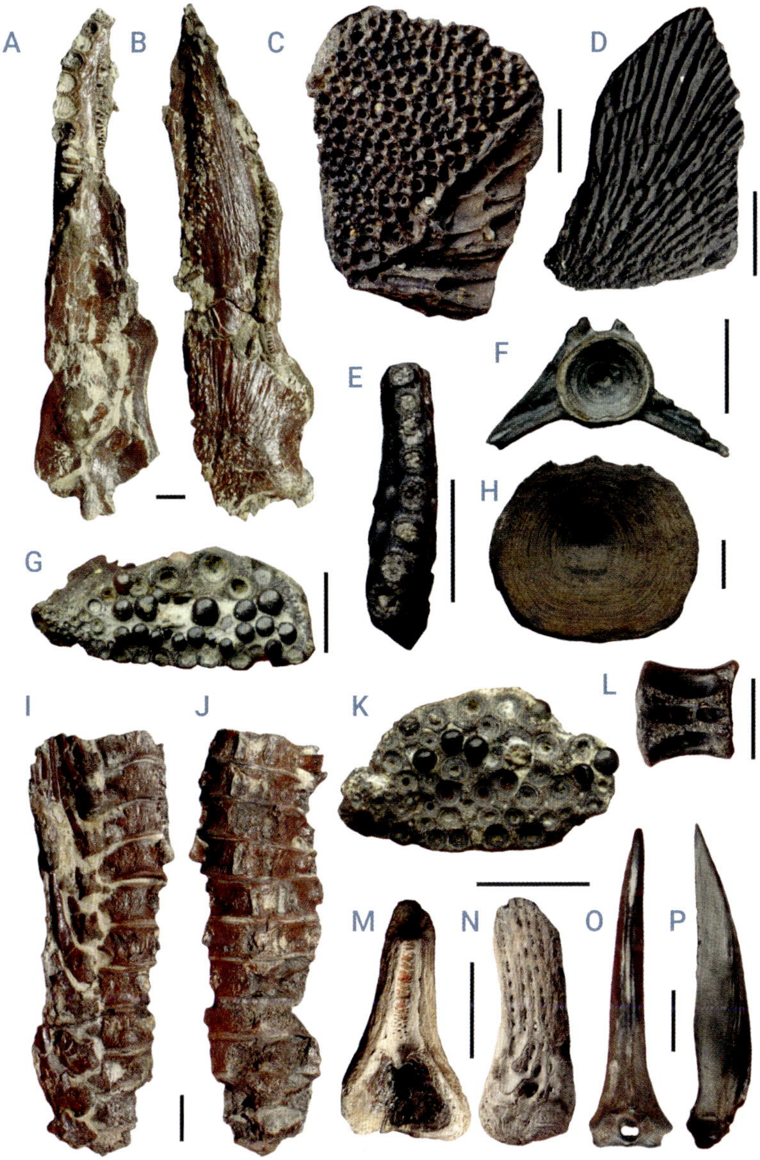

Figure 7.92

Fish fossils. A,B, partial bowfin skull. C, sturgeon plate. D, fragment of boney fish skull. E, ?*Amia* jaw. F, anterior view of a vertebra. G, K, durophagous fish tooth palette. H, ?*Amia* vertebra. I, J, series of Gar vertebrae with neural spines. L, lateral view of a fish vertebra. M, N, catfish fin spine. O, P, fin spine. A, B, I, J, collection of Theo Vickers. C, M, N, collection of Joe Thompson. D, E, F, H, K, L, O, P, collection of Rick Lilley and Charlotte Willis. Scale equals 10 mm

along the spinal column, they can possess a circular process on the sides, which is the rib attachment point. The scales are small shiny diamonds, usually black or dark brown, and can superficially resemble teeth due to the enamel coating on the scales. Jaw fragments are elongate pieces of bone with circular to square tooth sockets. In most instances, the teeth have been broken off. The skull bones have a unique ridged surface texture and are very shiny. The fin spines are tall, thin slightly curved pieces of bone that flare out

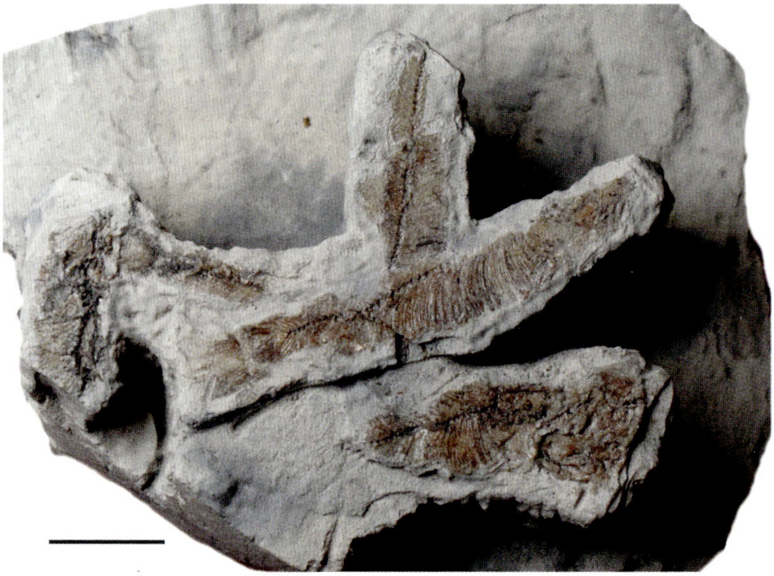

Figure 7.93

Paralates chapelcorneri from the Chapelcorner fish bed. Collection of Jack Wonfor. Scale equals 10 mm.

to a base with paired roller joint surfaces, and a foramen just above the base. Highly ornamented fin spines probably come from catfish (*Fig. 7.92*).

Chapelcorner fish bed.

The Chapelcorner fish bed is the islands own Konservat-Lagerstätten, meaning a site of exceptional preservation. The fish bed shales often yield a small herring-like fish called *Paralates chapelcorneri* and a prawn called *Propalaemon*. The fish occur fully articulated, with scales, fins, and sometimes even soft tissues preserved (*Fig. 7.93*). Isolated bones of the much larger predatory fish, *Amia* are also found commonly in the fish bed. The Chapelcorner fish bed is exposed at 3 locations, Kings Quay, Binstead and St. Helens, but the exact horizons are kept secret to protect the sites from overzealous collectors. In situ collecting at Kings Quay is not permitted as the site is protected.

Cartilaginous fish

The two known rays, *Myliobatus* and *Aetobatus*, that occur on the island can be identified by the shape and ornamentation of their tooth palettes. These are rows of long teeth which are modified to form a flat surface for crushing hard prey such as molluscs and

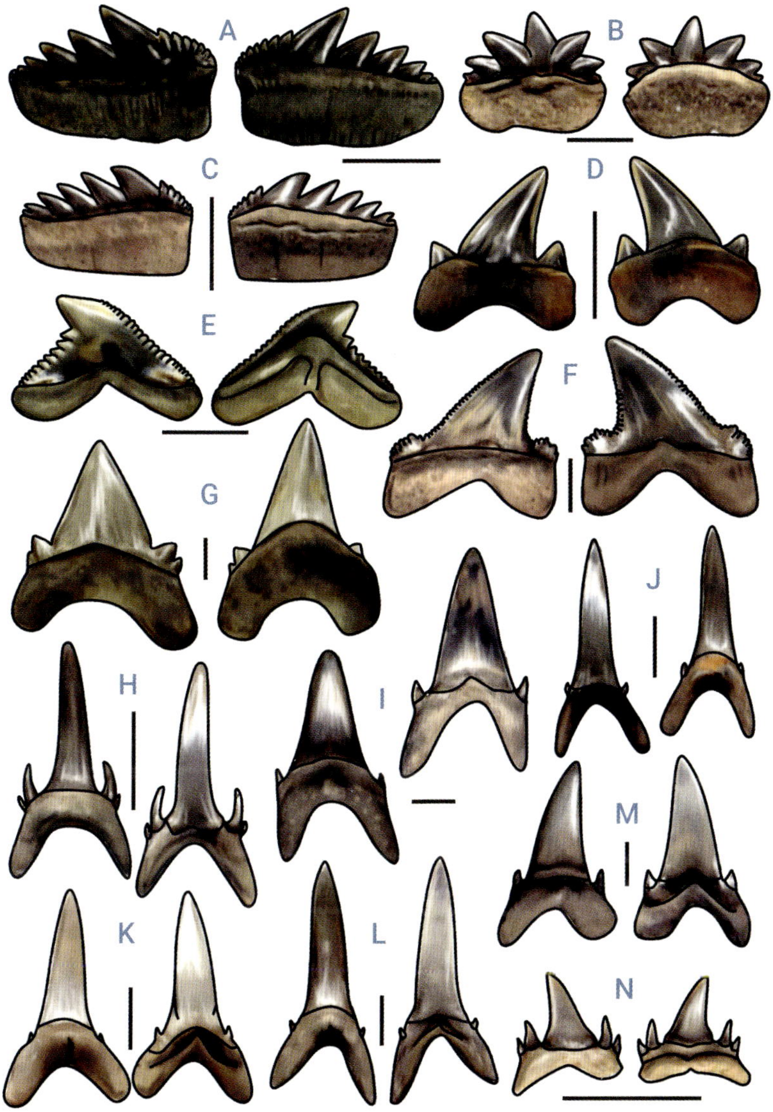

Figure 7.94

Digital drawings of common shark teeth.

A, *Notorhynchus serratissimus* anterior tooth. B, *Notorhynchus kempi* lower median symphyseal tooth. C, *Notorhynchus kempi* lower anterolateral tooth. D, *Cretalamina* sp. lateral tooth. E, *Galeocerdo latidens* lateral tooth. F, *Carcharocles auriculatus* lateral tooth. G, *Otodus obliqus* anterior tooth. H, *Carcharias acutissima* anterior tooth. I, *Jaekelotodus trigonalis* anterior tooth. J, *Hypotodus verticalis* anterior tooth. K, *Carcharias qustromensis* anterior tooth. L, *Striatolamia macrota* anterior tooth. M, *Hypotodus atlanticus* anterior tooth. N, *Carcharias acutissima* upper lateral tooth. Scale equals 10 mm.

crustaceans. Almost complete palettes can be found, but it is more common to find isolated shed teeth. Stingray tail spines are also found, being flattened elongate pieces of bone with a striated pattern on the surface. These have sharp barbs running along the sides, with their points directed towards the base of the spine.

Shark and ray vertebrae can resemble those of bony fish. However, the noticeable differences for the Isle of Wight faunas include the size, with shark/ray vertebrae being noticeably bigger,

Figure 7.95

Shark and ray remains.

Shark vertebra in A, lateral and B, anterior view. C, D, E, F, sections of shed ray teeth. A, B, collection of Rick Lilley and Charlotte Willis. Scale equals 10 mm.

and the vertebrae have two holes in the top of the vertebrae where cartilaginous processes once attached.

Shark teeth are only found at several localities on the north coast of the island, coming from the marine to brackish strata. They range in colour from being bluey black to brown, with a triangular crown and a V shaped root. The shark fauna is remarkably diverse, with the most common genera being *Carcharias*, *Hypotodus*, *Jaekelotodus Notorynchus*, *Odontaspis*, *Rhizoprionodon* and *Straitolamina* (*Fig. 7.94, 7.95*). The teeth range from millimetres across to a couple of centimetres.

Turtles

Fossil turtle shell is a common find on the north coast of the Isle of Wight. They come from our Late Eocene/Early Oligocene Solent Group clays and can be found along with the fossils of fish, alligators, birds, and mammals.

The turtle genera *Emys* and *Trionyx* that occur commonly as fossils are still alive today. *Emys* is known as the European pond

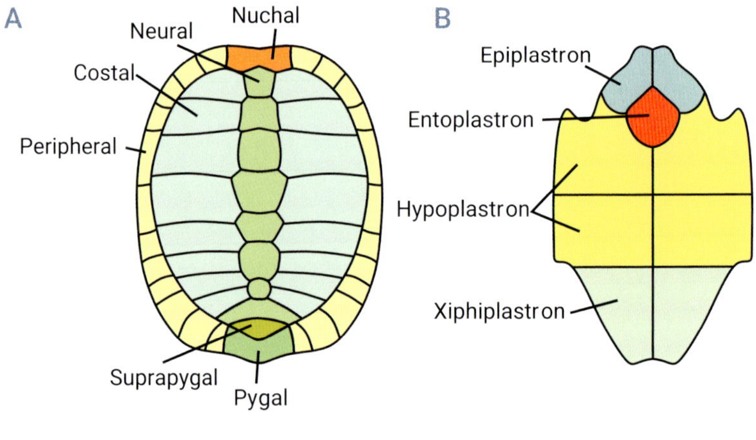

Figure 7.96

Diagram of a turtle shell anatomy.

Figure 7.97

Examples of turtle shell fragments.

A, pathological tortoise shell with parallel ridged ornamentation. B, partial *Trionyx* nuchal plate.C, almost complete peripheral plate of *Trionyx*. D, *Emys* neural plate in ventral view. E, *Emys* neural plate with bite marks. F, *Emys* peripheral plate showing scute margins. G, lateral view of a section of turtle shell showing the internal bone structure. H, I, peripheral plate with healed bite marks. J, *Trionyx* costal plate showing fused rib. B, C, D, collection of Rick Lilley and Charlotte Willis. F, H, I, collection of Sam Caethoven. J, collection of Theo Vickers. Scale equals 10 mm.

turtle, and *Trionyx* are known as African softshell turtles. They are easily distinguished by their surface texture, where *Emys* has a completely smooth shell, while *Trionyx* has a wavy ridged texture on the outer surface (*Fig. 7.96, 7.97*). Other skeletal elements, such as limb bones also occur, although not as commonly as the shell pieces (*Fig. 7.98*). Complete carapaces are also found by local collectors; turtle skulls are extremely rare.

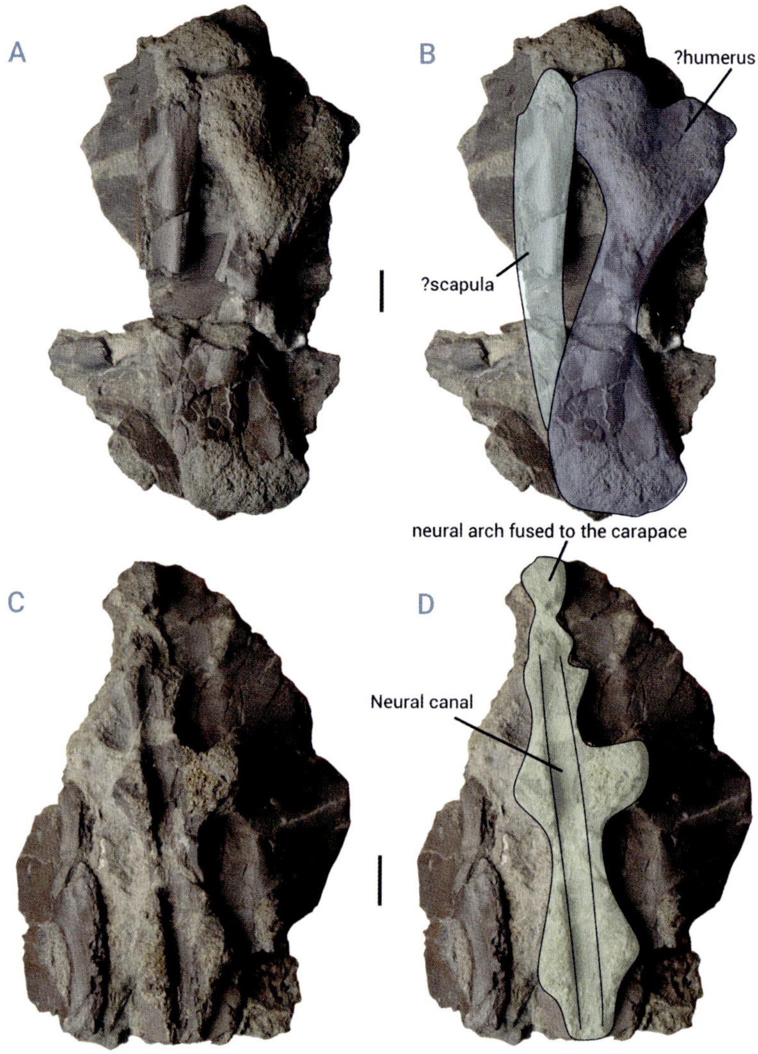

Figure 7.98

A, Section of *Trionyx* shell with a ?humerus and ?scapula. B, section of carapace with fused neural arches. Collection of Rick Lilley and Charlotte Willis. Scale equals 10 mm.

Alligator

The fossil alligator found on the island is called *Diplocynodon* hantoniensis, although it's likely with more complete finds that other species could be identified. *Diplocynodon* looked like the modern caiman alligator, reaching a maximum of 1.2 m in length. Remains of *Diplocynodon* are common, with teeth and scutes being the most recognisable. However, skull fragments, including jaws and vertebrae can also be found loose on the foreshore. Alligator osteoderms and skull bones are easily identified by a unique pattern of circular depressions over the outer surface (*Fig. 7.99–7.101*).

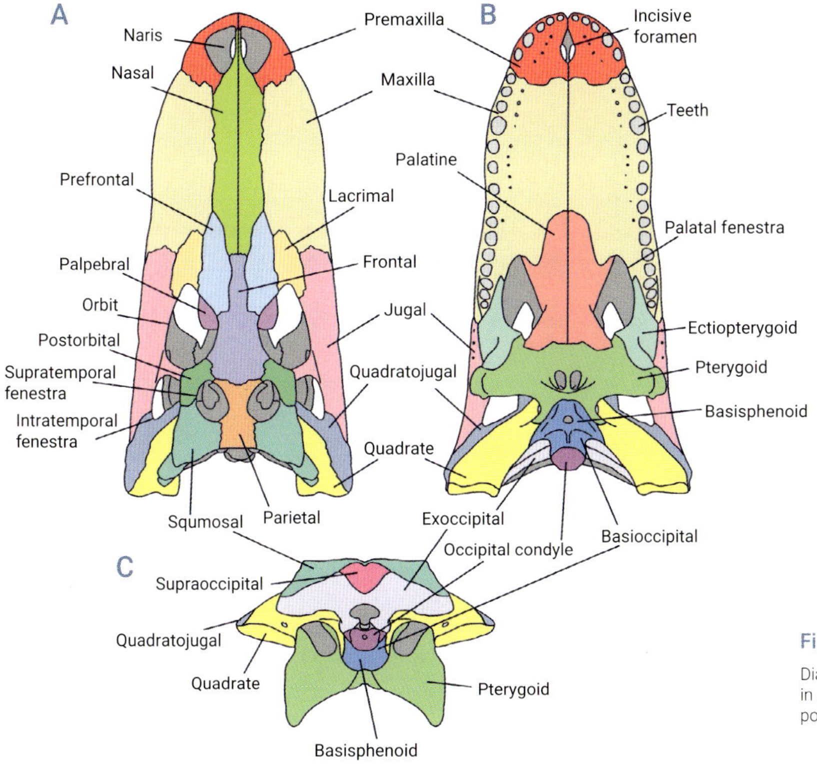

Figure 7.99

Diagram of an alligator skull in A dorsal, B, ventral and C, posterior views.

Snakes

Snake vertebrae also occur somewhat commonly and are easy to identify. All snake vertebrae have ball and socket articulation surfaces on either end of the centrum, as well as possessing 2 pairs of pre- and post- zygopophyses. The extra pair of zygopophyses aid in locking the vertebrae together for strength and reducing movement (*Fig. 7.108*).

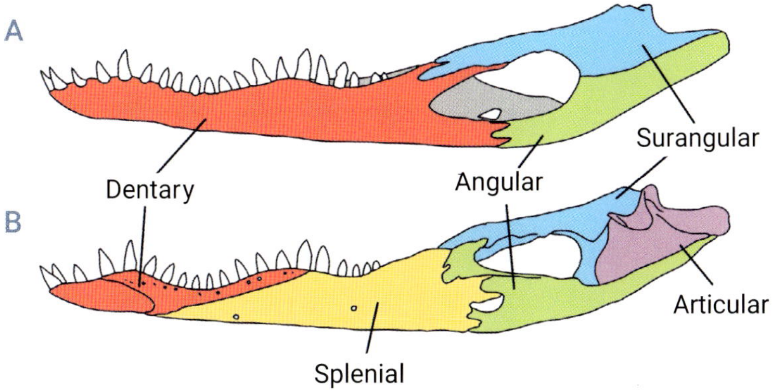

Figure 7.100

Diagram of an alligator lower jaw in A, lateral and B, lingual views.

Figure 7.101

Diplocynodon fossils.

A, dorsal view of a partial skull, showing parietals, frontals, prefrontals and nasals. B, fused parietals. C, dorsal osteoderm. D, ventral view of a partial skull. E, jugal. F, lateral view of a dentary. G, ?lateral osteoderm. H. nuchal osteoderm. I, J, tooth. K, ?lateral osteoderm. M, pair of fused dorsal osteoderms. A, D, F, L, collection of Theo Vickers. C, M, collection of Adie Butler. B, H, collection of Joe Thompson. E, G, collection of Sam Caethoven. K, collection of Rick Lilley and Charlotte Willis. Scale equals 10 mm.

Mammals

The Eocene and Oligocene strata on the island is of global importance for its mammal fauna, with several species known only from the island. The largest mammals are *Bothriodon* and *Palaeotherium*, easily identified by their distinctive teeth. Isolated elements of limbs, vertebrae and skull fragments are also frequently found (*Fig. 7.102–7.106*)

There is also a very diverse micro mammal fauna, however, to find these fossils involves collecting sediment samples and sieving

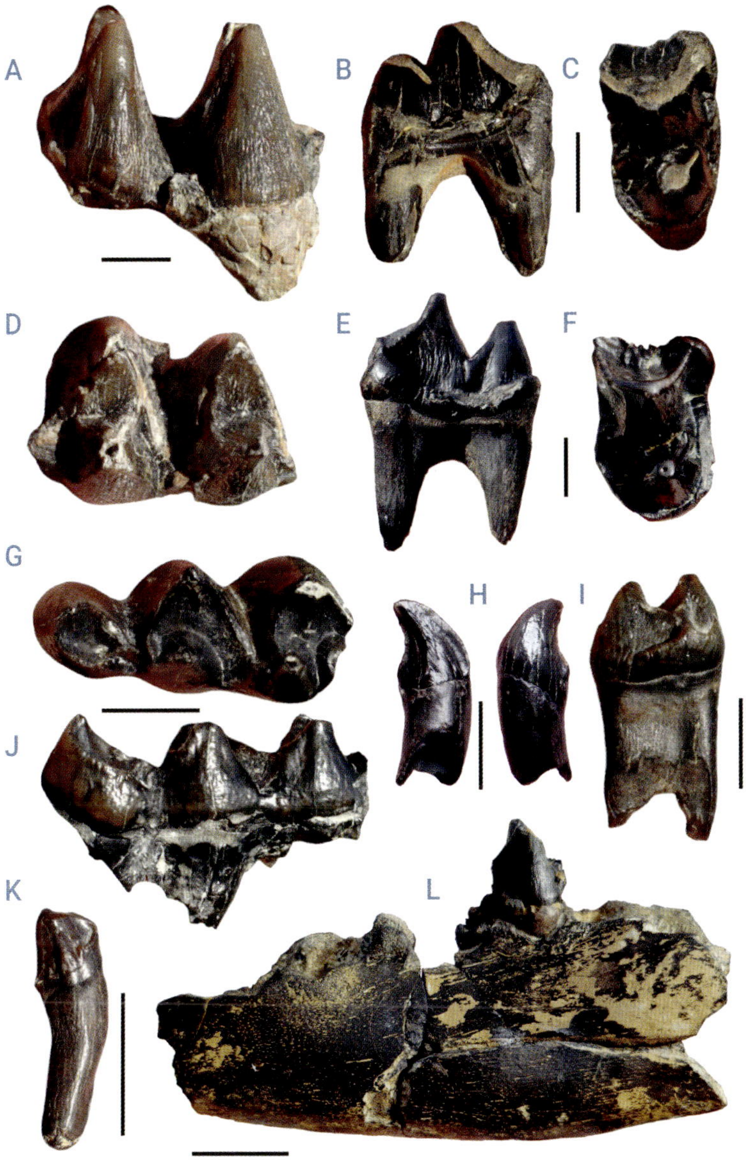

Figure 7.102

Mammal teeth.

A, D, *Bothriodon* lower molar. B, C, partial *Bothriodon* upper premolar. E, F, partial *Bothriodon* upper premolar. G, J, *Anoplotherium* lower third molar. H, unidentified mammal canine. I, Bothriodontine lower molar. K, bothriodontine anterior tooth. L, unidentified carnivoran jaw. A, B, C, D, G, I, J, K, collection of Theo Vickers. H, collection of Sam Caethoven. E, F, L, collection of Joe Thompson. Scale equals 10 mm.

for fossils under 1 mm. A binocular microscope is needed to see the fossils in the sieved residues.

Bone gnawing

On some fossil bones, small gnaw marks can be seen. Best viewed under a microscope, these are the traces of ancient rodents, which gnawed on the bones to help grind down and sharpen their ever-growing upper and lower incisors (*Fig. 7.107*).

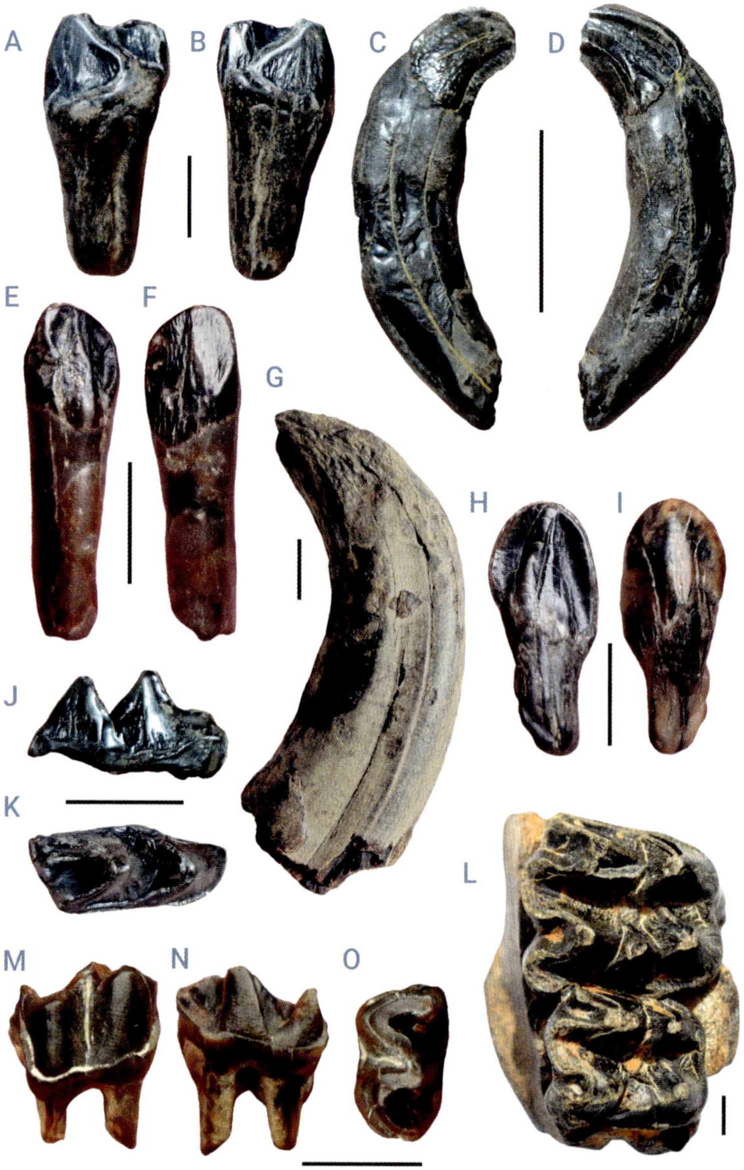

Figure 7.103

Mammal teeth.

A, B, partial *Bothriodon* lower molar. C, D, bothriodontine anthracothere incisor/canine. E, F, ?bothriodontid incisor. G, bothriodontine anthracothere canine root. H, I, partial bothriodontine anthracothere incisor. J, K, partial bothriodontine anthracothere upper molar crown. L, *Bothriodon* maxilla. M, N, O, *Paleotheirum* sp.. lower molar. A, B, E, F, H, I, J, K, collection of Sam Caethoven. C, D, G, collection of Joe Thompson. L, collection of Adie Butler. M, N, O, collection of Rick Lilley and Charlotte Willis. Scale equals 10 mm.

Grande Coupure event

The Grande Coupure event, otherwise known as the Eocene-Oligocene extinction event, occurred between 33.9 – 33.4 million years ago. This period saw a large-scale extinction of both flora and fauna, with many European Eocene mammals becoming extinct and new taxa from Asia appearing over a relatively short period. Pre Grande-Coupe, the island's mammalian fauna was dominated by

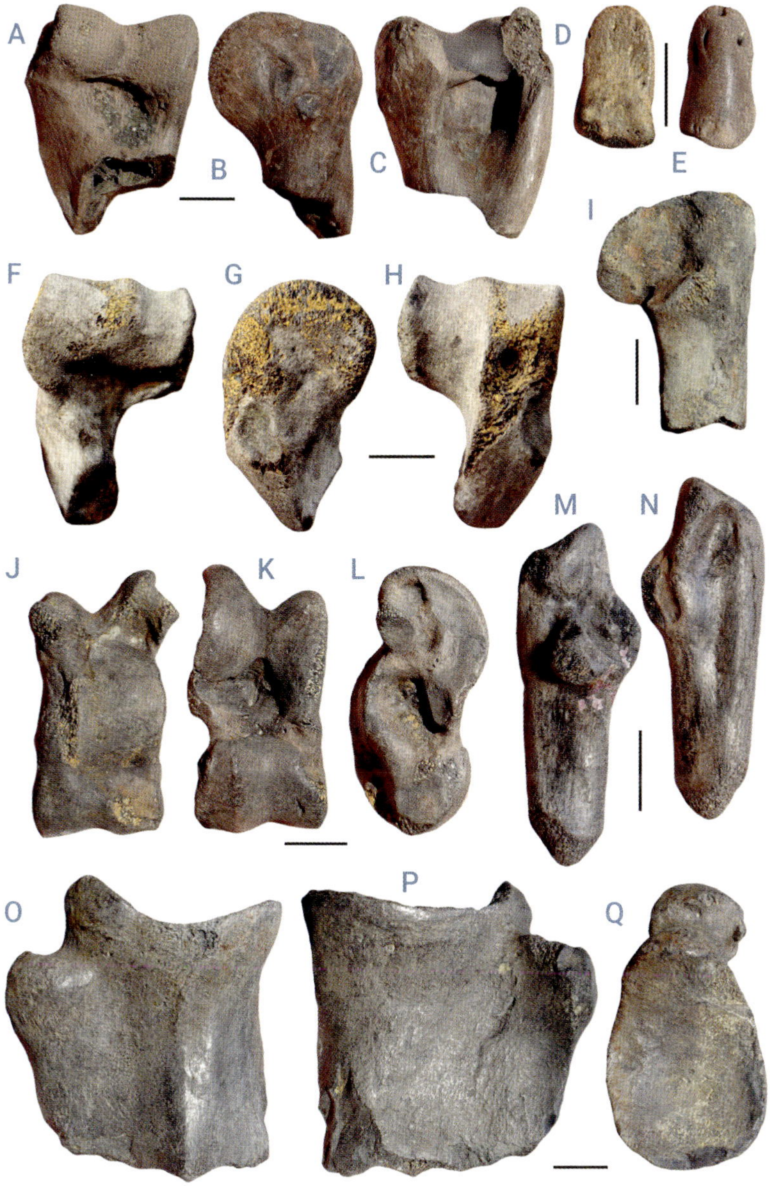

Figure 7.104

Limb bones.

A, B, C, distal *Paleotherium* humerus. D, E, *Bothriodon* ungual. F, G, H, *Bothriodon* distal humerus. I, ?*Paleotherium* calcaneum. J, K, L, *Bothriodon* astragalus. M, N, *Bothriodon* calcaneum. O, P, Q, *Ronzotherium* (primitive rhino) scapula (proximal end). A-E, I-L, M-Q, collection of Theo Vickers. F- H, collection of Joe Thompson. Scale equals 10 mm.

perissodactyls such as *Palaeotherium* (distant relative of the horse), a wide diversity of cloven-hoofed mammals, rodents and primates.

Post-Grande Coupe, saw the arrival of the families Rhinocertotidae (rhinos), Entelodontidae (pigs), Antracotheriidae (hippos), Cricetidae (hamsters, voles, lemmings), Castoridae (beavers) and Erinaceidae (hedgehogs) amongst others. Whilst

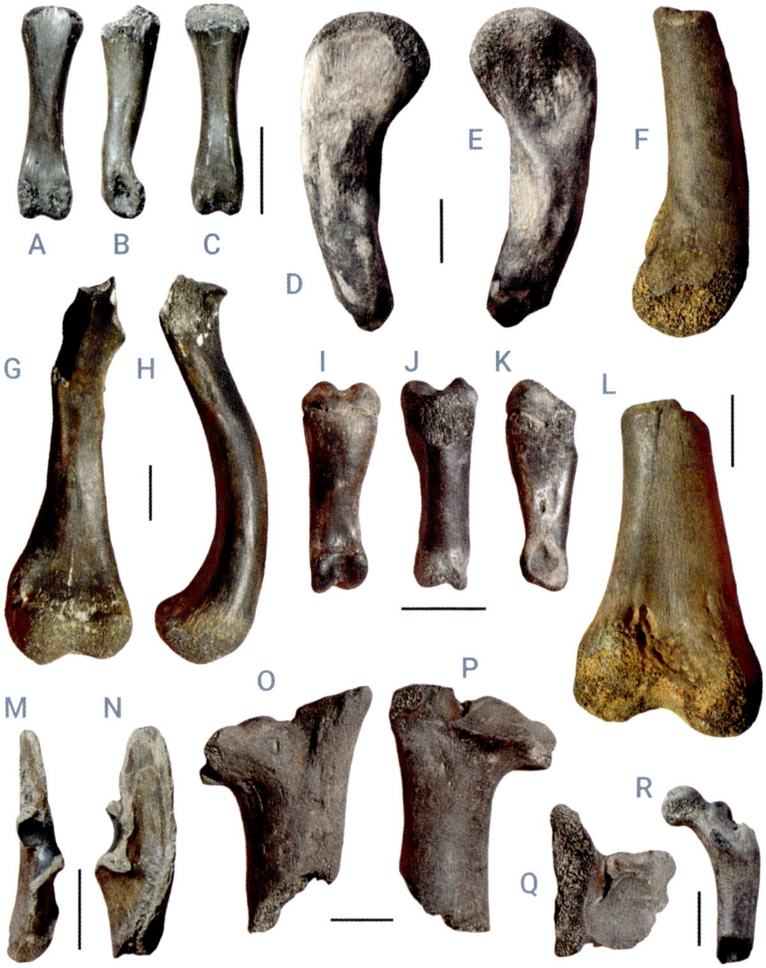

Figure 7.105

Limb bones.

A, B, C, *Diplocynodon* phalange. D, E, *Diplocynodon* femur (proximal end). F, K, *Diplocynodon* femur (distal end). G, H, *Trionyx* femur. I, J, K, *Bothriodon* phalange. M, N, unidentified mammal radius (proximal end). O, P, Q, ?*Anoplotherium* calcaneum. R, *Cynodictis* carnivorian mammal femur (proximal end). A-C, M,N, collection of Sam Caethoven. F, G-L, O-Q, collection of Theo Vickers. D, E, R, collection of Joe Thompson. Scale equals 10 mm.

Palaeotherium, *Anoplotherium* and the families Xiphodontidae (even-toed ungulates) and Amphimerycidae (relatives of camels and llamas) and almost all European primates disappeared completely.

Cause?

There is global evidence for abrupt climatic cooling during this period, called the Oi-1 Event, with $\delta^{18}O$ (ratio of stable oxygen isotopes, ^{18}O and ^{16}O) values falling dramatically at 33.7Ma. This decrease in $\delta^{18}O$ values corresponds to the development of the Antarctica ice sheet, with widespread global temperature falls at the same time.

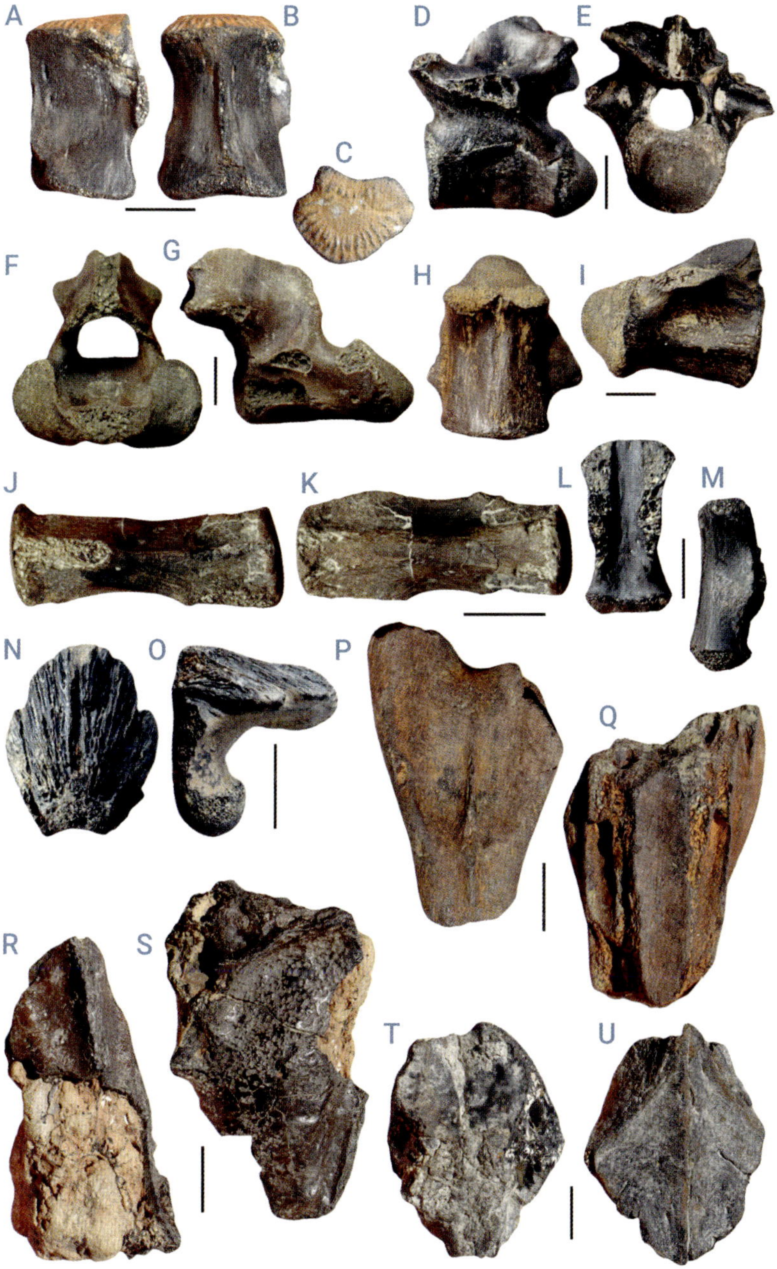

Figure 7.106

Vertebrae and skeletal material.

A, B, C, bothriodontine anthracothere vertebra. D, E, *Diplocynodon* dorsal vertebra. F, G, turtle vertebra. H, I, *Diplocynodon* cervical vertebra. J, K, *Diplocynodon* caudal vertebra. L, M, mammal caudal vertebra. N, O, *Diplocynodon* occipital condyle. P, Q, *Bothriodon* symphesium (lower jaw). R, S, anoplotheriid mammal partial skull. T, U, *Bothriodon* skull roof. A, B, O - T, collection of Theo Vickers. C, D, K- N, collection of Joe Thompson. E, F, I, J, collection of Rick Lilley and Charlotte Willis. Scale equals 10 mm.

It is also thought that the dispersal of animals from outside the mainland European landmass played a role in the mammal extinctions. Asian mammals expanding their range simply outcompeted the European mammals, leading to their demise.

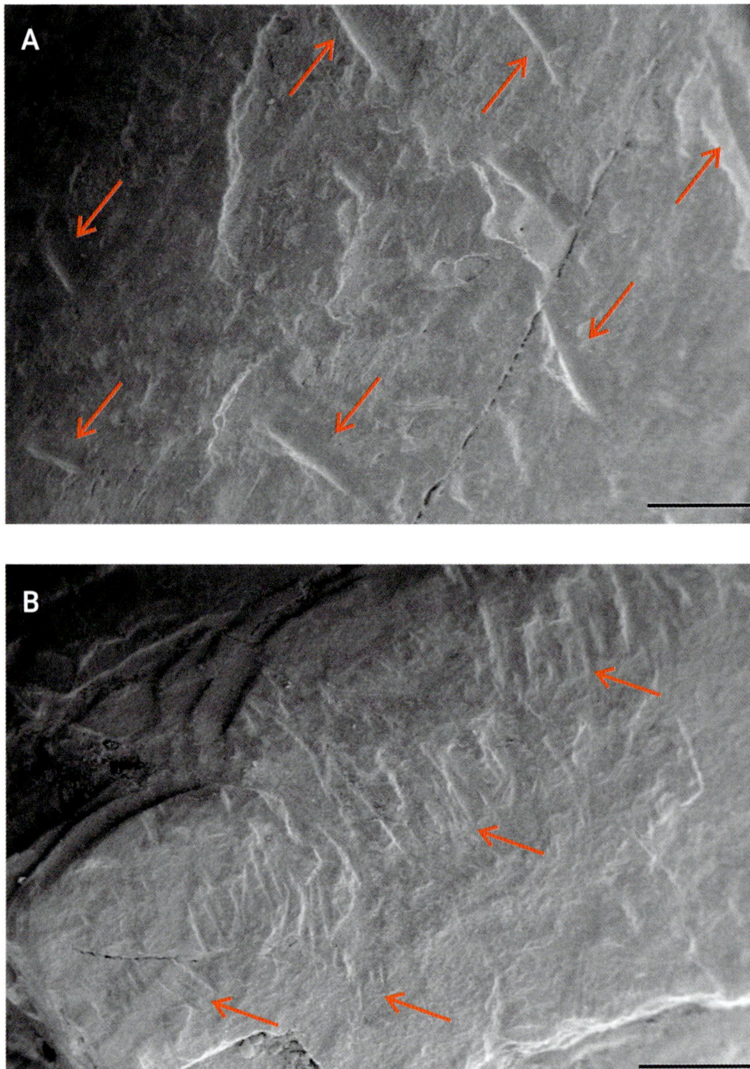

Figure 7.107

Scanning electron microscope (SEM) photos of bone fragments showing gnaw marks made by small rodents. Scale equals 1 mm.

Birds

The remains of fossil birds on the island are incredibly rare, despite being well-known from equivalent rocks on the UK mainland and elsewhere in Europe. However, isolated bones do turn up from time to time and are distinct from the usual alligator, turtle and mammal bones. Firstly, most bird bones are hollow, with a thin bone wall. They have highly distinctive shapes allowing experts to identify down to species level to which bird they belong to. One of the most

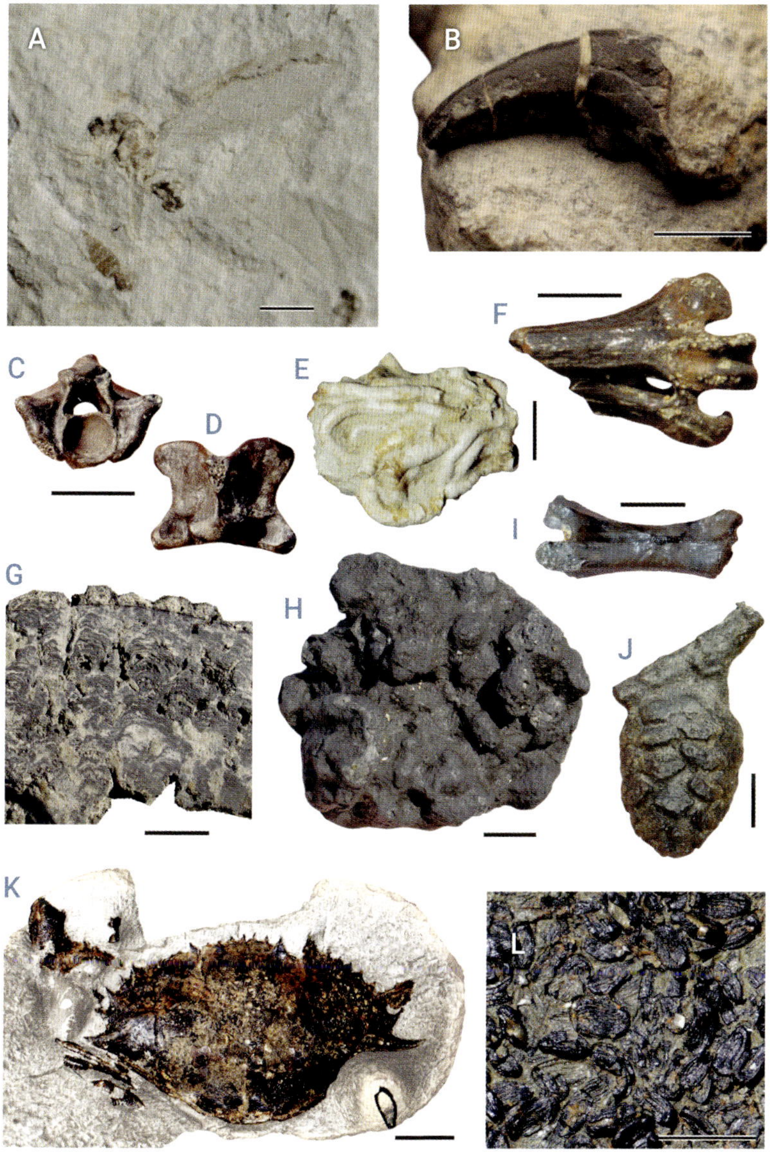

Figure 7.108

Other Eocene fossils.

A, winged insect and ant from the insect limestone. B, crab claw. C, D, snake vertebra in posterior and dorsal views. E, serpulid worm cluster. F, bird tarsometatarsus. G, cut and polished stromatolite. H, stromatolite. I, lizard caudal vertebrae. J, pyritised pinecone. K, *Neptunus vectensis* crab from the Hamstead. L, *Stratiotes*, Water soldier seeds. A, F, collection of Theo Vickers. B, collection of Rick Lilley and Charlotte Willis. C, D, I, collection of Sam Caethoven. J, L, photos courtesy of Adie Butler. K, found by Theo Vickers, photo courtesy of John Quayle. A, scale equals1 mm. B, C, D, E, F, I, L, scale equals 5 mm. G, H, J, K, scale equals 10 mm.

distinctive bird bones is the tarsometatarsus. This is the 3 metatarsal bones fused into one, in dinosaurs, these are usually 3 separate bones (*Fig. 7.108*).

The Eocene bird fauna on the island closely resembles what we have today but also globally, with a range of waders, ducks and owls, along with the ancestors of flamingos and grebes, gannets, boobies, loons, potoos, swifts and hummingbirds.

Flora

The tropical conditions during the Eocene – Oligocene provided excellent conditions for plant growth, with the environment likely resembling the Florida Everglades. Sequoia twigs, seeds and palm fronds are the most common finds, and a coal seam is present at Alum Bay within the Bracklesham Group (Marsh Farm Formation), but it is too thin to be of commercial value.

Other fossils

The Paleogene strata also yield countless other fossils, although slightly less commonly. Crustaceans are present throughout the formations, with lobsters and barnacles in the marine horizons only, and crabs found throughout. Other common finds are serpulid worms, bryozoans, and echinoids. The internal casts of insect cocoons (possibly wasps) are found commonly in the Bembridge Limestone (*Fig. 7.108*).

Insect Limestone

The preservation of insects is a very rare phenomenon, mostly occurring in finely laminated shales, limestones and amber. However, on the island we are lucky enough to have the so-called 'insect limestone'. This is a lensed deposit within the Bouldnor Formation, only outcropping at Bembridge and Thorness Bay and is an internationally important horizon yielding exquisitely preserved insects, shrimp, rare feathers and plant remains. It is thought it was deposited in a series of shallow lagoons, that were low energy with periods of rapid deposition of mud.

The insect limestone blocks are easily recognised as loose pebbles on the foreshore, being light bluey cream in colour and very fine-grained. Light taps with a geological hammer will split these pebbles, hopefully revealing insect fossils (*Fig. 7.108*). A hand lens is essential, as many are incredibly small and hard to see with the naked eye. Collecting at this locality is very hit-and-miss and requires considerable patience.

Stromatolites.

Stromatolites are formed by microbial communities of cyanobacteria (blue-green algae), trapping sediment or precipitating calcium carbonate to form large domes. Stromatolites were some of the first lifeforms on earth, appearing some 3 billion years ago, and becoming the dominant lifeform on the planet. Until the Holocene (11,700 years ago), stromatolites were widely distributed, but today they can only be found in a handful of places. Some colonies can grow over a metre across, and it's thought the largest colonies are 2-3,000 years old.

On the island, stromatolites can be found along the shores at Burnt Wood, coming from the base of the Gurnard Member. Here they can reach 20 cm across (*Fig. 7.108*).

Coprolites

Coprolites are very common along various stretches of the north coast of the Isle of Wight and are mostly assigned to alligator *Diplocynodon*. However, it's more likely they are produced from various mammals as well as the alligators (*Fig. 7.109*).

Figure 7.109

Examples of coprolites from the Bouldnor – Hamstead area. A, B, collection of Theo Vickers. C, collection of Joe Thompson. Scale equals 10 mm.

7.6. Palaeolithic to Holocene

With contributions by Theo Vickers.

The Pleistocene and Holocene epochs cover the period from 2.6 million years ago to today. This was a period dominated by repeated glaciations, frequent episodes when Earth's ice sheets significantly expanded and the climate dramatically cooled, giving rise to its infamous name the 'Ice Age'. These 'glacials' were also punctuated by warm episodes called 'interglacials' with temperatures as warm or warmer than today. This chaotic fluctuating climate is key to understanding the major changes to the landscapes, environments and ancient fauna of southern England and the Isle of Wight that took place during this period, all culminating in the modern island environment we see today.

For much of the Pleistocene or 'Ice Age', the Isle of Wight was a range of chalk hills and lowlands south of present-day Hampshire. The region was dominated by the large Solent River system and its tributaries, which we see today as the modern Solent, and the rivers that feed it such as the Yar, Medina, western Yar, Beaulieu, Hamble, Test and Itchen. During cold glacial periods, ice sheets expanded southwards and sea levels would fall, creating open tundra and steppe environments similar to modern Siberia or northern Canada, south of the ice sheets. During some larger glacial periods, the ice sheets could reach as far south as London and the Midlands, and areas south of this, including the Isle of Wight, would be uninhabitable glacial deserts.

Warm interglacials would see the landscape blanketed with balmy forests, grasslands and open woodlands, similar to those we see today. Megafauna such as extinct straight-tusked elephants, rhinos and hyenas inhabited the landscape, and high sea levels would cause estuaries, beaches, lagoons and coastal plains to form across our region. Britain likely remained connected to mainland Europe during these episodes, but very warm interglacials may have seen Britain and possibly the Isle of Wight become islands for the first time, particularly during the last interglacial c.120,000 years ago. As the climate fluctuated between these extremes for much of the Pleistocene, rivers, estuaries, wetlands and beaches laid down sequences of sediment on the Isle of Wight, particularly around the Solent River and its tributaries, leaving a record of these changes and preserving the remains of the Pleistocene megafauna that roamed these landscapes.

As humans began inhabiting the island, from nomadic tribes to permanent settlements, they brought domesticated breeds of cattle, horses, and sheep. Evidence of this recent prehistory is scattered across the island and mostly found as bones washed up loose on the beaches from outcrops just a short distance offshore. Gravels on the top of the cliffs on the southwest coast of the island also yield some of these more recent faunas.

Megafauna

The island's Pleistocene strata with its sequences of river gravels, raised beaches, sands, periglacial soils and intertidal muds can yield the fossils of a range of large 'Ice Age' mammals. These are the youngest fossils to be found on the island, and while generally quite rare, the fossils of species belonging to two distinct mammal faunas can be found, representing the fluctuating cold glacial and warm interglacial periods (*Fig. 7.110, 7.112*).

While the exact species present varied significantly throughout the Pleistocene, glacial periods in Ice Age Britain with their tundra and open steppe habitats supported faunas dominated by herds of woolly mammoth, reindeer, wild horses, steppe bison, musk ox and woolly rhinoceros. These herbivores were in turn preyed upon by a range of predators such as large cave hyenas, cave lions, wolves and later in the Pleistocene, well-equipped Neanderthals and the modern humans, *Homo sapiens*.

On the Isle of Wight, the fossils of glacial fauna typically occur within the river gravel sequences, most notably on the island's southwest coast. Here most finds are typically made after cliff falls, with fossils either loose on the shoreline or in fallen material at the base of the cliff. The vast majority of these finds consist of isolated material that was washed into river channels, and as a result can be reworked, abraded or incomplete, and often very fragile. On the

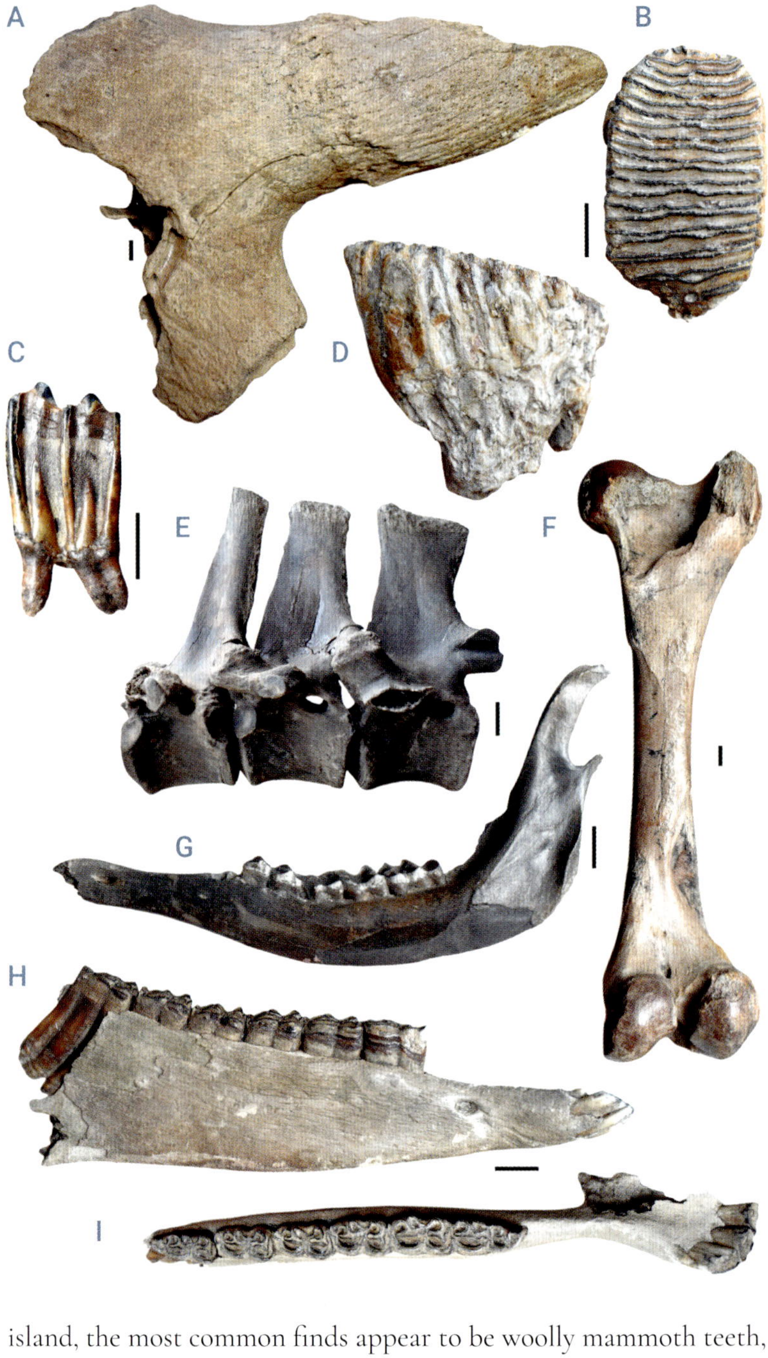

Figure 7.110

Ice Age mammal remains.

A, partial steppe bison skull showing the eye socket and horn. B, D, mammoth tooth. C, isolated steppe bison tooth. E, articulated steppe bison vertebrae. F, steppe bison femur. G, deer mandible. H, I, horse mandible. B, D, E, collection of Jack Wonfor. C, F-I, collection of Theo Vickers. Scale equals 10 mm.

island, the most common finds appear to be woolly mammoth teeth, bones and occasional tusk sections, alongside material from steppe bison, wild horses and more rarely reindeer and woolly rhinoceros.

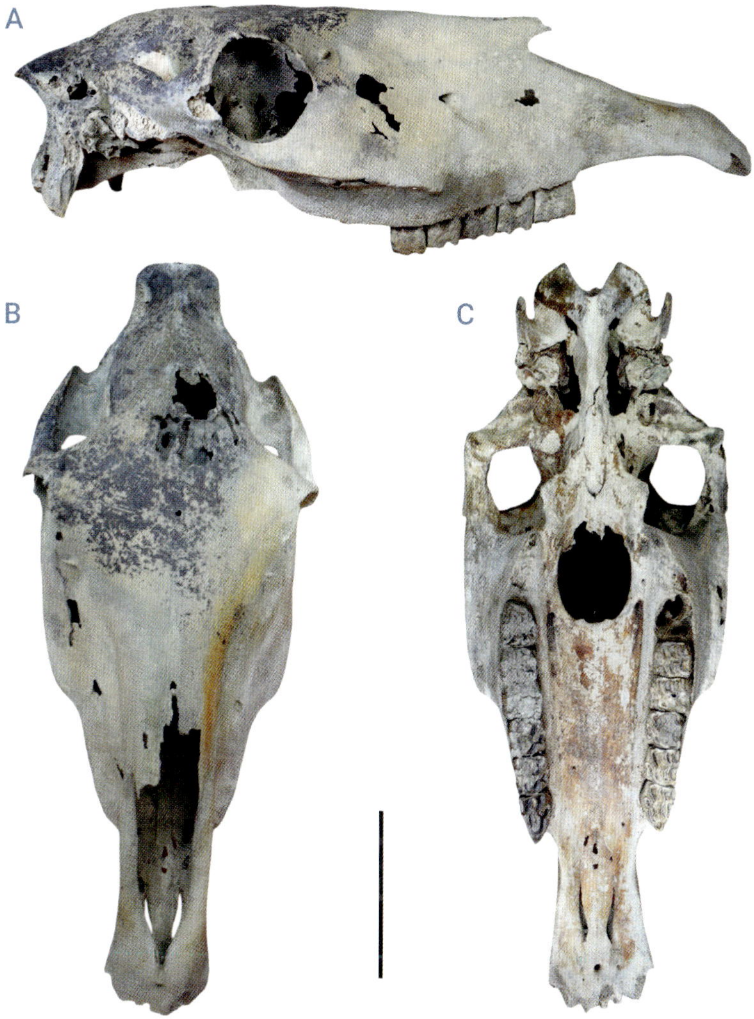

Figure 7.111

Horse skull found by Jack Wonfor. Scale equals 10 cm.

The exact age of these fossils can be unclear, but most likely date to milder phases during the last glacial period 12,000 - 115,000 years ago.

Warmer interglacial periods hosted quite different mammal faunas, consisting of many species we would be familiar with today including red, roe and fallow deer, wild boar, foxes and badgers. Alongside them however was an extensive array of megafauna, commonly including herds of extinct straight-tusked elephant (*Palaeoloxodon antiquus*), steppe bison, several species of rhinoceros,

Figure 7.112

Luke and Jo Ferguson with their impressive straight-tusked elephant humerus.

hippopotamus, giant deer, wild horses and the powerful aurochs, hunted by lions, hyenas, leopards, wolves, wild dogs and ancestral brown bears (*Fig. 7.112*). Interglacials also offered good conditions for humans, and the presence of species such as *Homo heidelbergensis* was probably limited to these warmer periods. Fossils of these interglacial mammals on the island occur in an array of deposits including intertidal muds and river gravels, particularly along the island's northern and to a lesser extent south-west coastline. Most finds are isolated bones, tusks, horns, antlers and teeth, from steppe bison, deer, straight-tusked elephant and aurochs. In intertidal mud deposits, fossils can be well-preserved with complete skulls and articulated remains of steppe bison and aurochs found. The fossils of interglacial mammals from the island's northern coastline can be

Figure 7.113

A, selection of Noah's nuts and B, a section of wood with the bark still preserved from the gravels at Brook Bay.

dated to the last interglacial period around 120,000 years ago, but some interglacial mammal fossils from the island's south-west coast may be from older deposits.

Holocene Forests at Brook and Compton Bay

Along the cliff line at Brook and Compton Bay on the island's southwest coast, sequences of Pleistocene gravel and Holocene peat and soil layers can be seen overlying the Cretaceous rocks. These deposits represent the lost headwaters of the western Yar, which once flowed north across a landscape to the valley at Freshwater Bay. Contained within the organic Holocene sediments are evidence to wooded landscapes that took hold in the early Holocene, particularly in the form of 'Noah's nuts'. Noah's nuts are ancient hazelnuts, dated to around 8000 years ago (*Fig. 7.113*), and have been noticed by islanders for over 200 years. They often have small circular holes on the surface, which are traces of ancient dormice and squirrels eating the nuts. These are commonly found when cliff falls bring sections of this 'nut bed' to the beach. Alongside the hazelnuts pieces of wood can be commonly found too. This wood is so young it's not fossilised and can burn easily once dried. The species are largely oak and hazel wood, giving glimpses of the ancient Holocene forests that covered the island's landscape after the tundras of the Pleistocene retreated. Large logs can often be seen protruding from the Pleistocene gravels at Brook Bay, representing older trees from this time. The bones of deer have also been found in the Holocene deposits, alongside Mesolithic flint tools.

7.7. Ancient humans

With contributions from Theo Vickers.

Flint Tools

By far the most common prehistoric archaeological finds on the island are flint tools. Referred to as 'lithics' by archaeologists, flint tools can be encountered quite commonly along the island's shoreline and also inland in ploughed fields, gardens and footpaths. On the island, flint tools can vary in age considerably from the lower Palaeolithic over 500,000 years ago, through to the early Bronze Age just 4,000 years ago. They represent communities throughout these distinct prehistoric periods producing a diverse array of tool types, reflecting their changing lifestyles, technology, environment and cultures. If found it is always worth presenting flint tools to experts at the island's local archaeology service or submitting them for recording in the Portable Antiquities Scheme, especially with rare or well-preserved artefacts.

Palaeolithic.

In Britain, the Palaeolithic represents a vast period of the Pleistocene, lasting from around 900,000 to just 11,700 years ago. On the Isle of Wight, the earliest Palaeolithic, and oldest flint tools we can encounter belong to a tool technology called the 'Acheulean', dating to 250,000 - 500,000 years ago. It was long thought that Acheulean tools were made by the earliest humans in Britain, but recent discoveries of basic flint tools and footprints in Norfolk now suggest an ancient human species, such as Homo antecessor, were present during warm periods as far as back as 850,000 - 900,000 years ago. Currently, no evidence of these earliest people has been found around the Solent, and Acheulean tools remain the oldest evidence of humans on the Isle of Wight. Lower Palaeolithic Acheulean flint tools often consist of hand axes, robust and crude scrapers and large flakes, which due to their age can have heavy mineral staining (referred to as 'patina') from the iron-rich sediments they come from. Hand axes in particular are recognisable and beautiful artefacts of this time and were likely used primarily as butchery tools. The makers and users of hand axes and Acheulean tools were probably an archaic human species we know as *Homo heidelbergensis*, and later on their probable descendants, early Neanderthals (*Homo neanderthalensis*). In general, these lower Palaeolithic tools are rare finds but have been found at various sites across the Isle of Wight, often along stretches of shoreline with eroding deposits of Pleistocene river gravels. Overall, they represent the most common artefacts to be found on the island from the whole span of the Palaeolithic.

Middle Palaeolithic flint tools, referred to as 'Mousterian', have also been found at several sites on the island, and typically date from around 45,000 - 200,000 years ago. These Middle Palaeolithic 'Mousterian' tools are again rare find on the Isle of Wight but can include artefacts such as large scrapers, cutting tools, spear points (produced using a specific technique called 'Levallois') and 'bout-coupe' hand axes used for butchery. Like lower Palaeolithic artefacts they can have heavy patinas due to their age and are often fairly large. In Britain, these tools were made by Neanderthals, who would expand north into Britain whenever climate and sea levels allowed them to. For much of the mid to late Pleistocene however, Britain appears to have been uninhabited by humans, in part due to severe cold periods, but also the particularly high sea levels quickly cutting Britain off from mainland Europe as an island. Meaning that while habitable, Britain was inaccessible to Neanderthals unwilling or unable to cross the wide stretches of water.

45,000 - 50,000 years ago early populations of our species (Homo sapiens) began to appear in Europe, marking the start of the Upper Palaeolithic, and after a brief period of co-existence, the end of the Neanderthals. These modern human hunter-gatherers reached southern England around 45,000 years ago, but their numbers were extremely small and their presence sporadic, meaning evidence of them is limited to a handful of rare sites across Britain. Severe glacial conditions from 35,000 years ago onward then followed, making Britain uninhabitable for thousands of years with conditions only starting to improve around 15,000 years ago as the last Ice Age began to thaw. Modern hunter-gatherers returned to Britain soon after this time, and it seems these later Upper Palaeolithic people were active in the Solent region, with scattered flint tools found in southern Hampshire, and a particularly large site at Hengistbury Head in east Dorset. To date, however, no early or late Upper Palaeolithic flint tools have ever been identified on the Isle of Wight. Despite this, the presence of late Upper Palaeolithic activity in the region means that while likely extremely rare, Upper Palaeolithic flint artefacts, often consisting of long blades, delicate points and scrapers, may plausibly be found on the Isle of Wight (*Fig. 7.114*), and any possible finds should be presented to local experts.

Figure 7.114

Palaeolithic flint tools.

A, E, Lower Palaeolithic hand axe. B, C, Lower/Middle Palaeolithic Flake or Cutting Tool. D, F, Lower Palaeolithic hand axe. A, B collection of Andrew Jacobs. D, F, collection of Nathan Skene. Scale equals 10 mm.

Mesolithic.

11,700 years ago, the climate started to warm, causing rising sea levels and an increase in woodland habitats, along with a change in fauna, marking the beginning of the Mesolithic. Due to the environmental change, hunter-gatherers had to adapt, which is evident in the flint tools being produced.

Mesolithic flint tools, broadly dating from 6,000 - 10,000 years ago, can be found regularly on the Isle of Wight, particularly along the shore of the island's northern Solent coastline, and are typically

Figure 7.115

Neolithic and Mesolithic flint tools.

A, Late Mesolithic Blade. B, broken Early Neolithic Axe. C, Late Neolithic/Early Bronze Age Plano-convex 'slug' Knife. D,?Neolithic pick. E, Mesolithic blade. F, Mesolithic adze. G, Late Neolithic/Early Bronze Age Scraper. B, collection of Adie Butler. C, F, G, collection of Theo Vickers. D,F, collection of Andrew Jacobs. Scale equals 10 mm.

fine bladelets, blades, blade cores, picks, and adzes. Microliths, very small and well-made blades and points used in objects such as large arrows, are particularly diagnostic Mesolithic finds with different forms dating to different stages of the Mesolithic. Axe-like adzes are another characteristic tool used for woodworking and have been found on the island's northern coast. The raw flint used to produce these Mesolithic implements is often of varied origin and in some cases may not be from the Isle of Wight, representing the mobility of Mesolithic people. The patina of Mesolithic material on the island can vary, but when found on the intertidal shoreline can be a dark blue/grey to glassy black (*Fig. 7.115*).

One of the island's (and the world's) best Mesolithic sites lies below the Solent itself off Bouldnor Cliff, where maritime archaeologists have uncovered an internationally important and well-preserved Mesolithic camp eroding on the seafloor at a depth of 11m. Dating to around 8,000 years ago, the site preserves examples of worked wood, flint tools and small occupational structures, alongside evidence for the production of Mesolithic boats. At the time, the Solent was a wide low-lying river valley home to rich woodlands and wetlands, full of resources for the Bouldnor Cliff community to exploit. Rising sea levels both during and after occupation at Bouldnor Cliff led to the flooding of the western Solent by the sea, separating the Isle of Wight from mainland Britain around 7000 years ago and drowning this Mesolithic landscape below the waves.

Neolithic and Early Bronze Age.

Beginning around 6000 years ago prehistoric Britain underwent a significant cultural shift. Neolithic farming communities from what is today northern and western France set out across the English Channel to settle in Britain and Ireland. Mesolithic hunter-gatherers disappeared, and the flint toolkit we find changed as farmers and their new way of life took hold of the landscape, clearing and burning the forests and constructing farmsteads and monuments. On the Isle of Wight, flint tools from the Neolithic are reasonably common finds both around the coast and inland, and along with material from the early Bronze Age are the most common flint tools to be encountered on the island. The island's Neolithic communities likely lived in small, scattered farmsteads or settlements, with an economy based on farming livestock and crops. The most common early Neolithic flint finds are debitage, the debris of flint tool making, including struck flakes, chips and blades, but tools such as scrapers, knives and piercers can be regularly found, alongside early pottery. Rarer early Neolithic finds on the island can include polished axes and leaf arrowheads.

Some polished axes are made of flint and were probably used as tools/weapons, but others from the island were made of exotic non-local rocks such as basalt and dolerite and were probably directly imported or traded as prestigious items of cultural/ritual significance (*Fig. 7.116*). Aside from this, much of the island's Neolithic and later early Bronze Age flint tools are made from local flint taken from deposits of Pleistocene river gravel, beach cobbles, pebbles and probably from the chalk itself. The patina of Neolithic to early Bronze Age material on the island is highly variable and ranges from whitish blue in chalk-rich soils to glassy grey, red, yellow and fresh black depending on where artefacts were buried (*Fig. 7.116, 7.117*).

Around 4,500 years ago, during the heyday of ritual monuments like Stonehenge, a new wave of migrants began to appear in late Neolithic Britain. These were people from the area of modern Belgium, Germany and the Netherlands, whose ancestors had been part of a large migration from the Eurasian steppe. We call these new arrivals the 'Beaker People', after the distinctive type of decorated pottery they used (*Fig. 7.118*), and they appear to have introduced significant cultural changes, including metal-working, to Britain and initiated the transition to the early Bronze Age. While metal tools and weapons were now being made, however, these were probably prestige items and for most people flint tools remained a part of everyday life.

On the Isle of Wight later Neolithic/early Bronze Age flint finds are widespread and seemingly more common than those from the earlier Neolithic. Finds typically include discoidal, end and side scrapers, debitage, cores and 'slug' knives, along with rare beaker pottery.

Beautiful barbed and tanged arrowheads are particularly characteristic finds from this 'beaker period' and have been found across the island (*Fig. 7.117*). Some were used for their intended purpose, but others were often left (probably hafted to arrows) as grave goods alongside beaker pots. As Bronze Age society continued to develop from 4,000 years ago onwards flint working soon declined in Britain as bronze tools became more common and

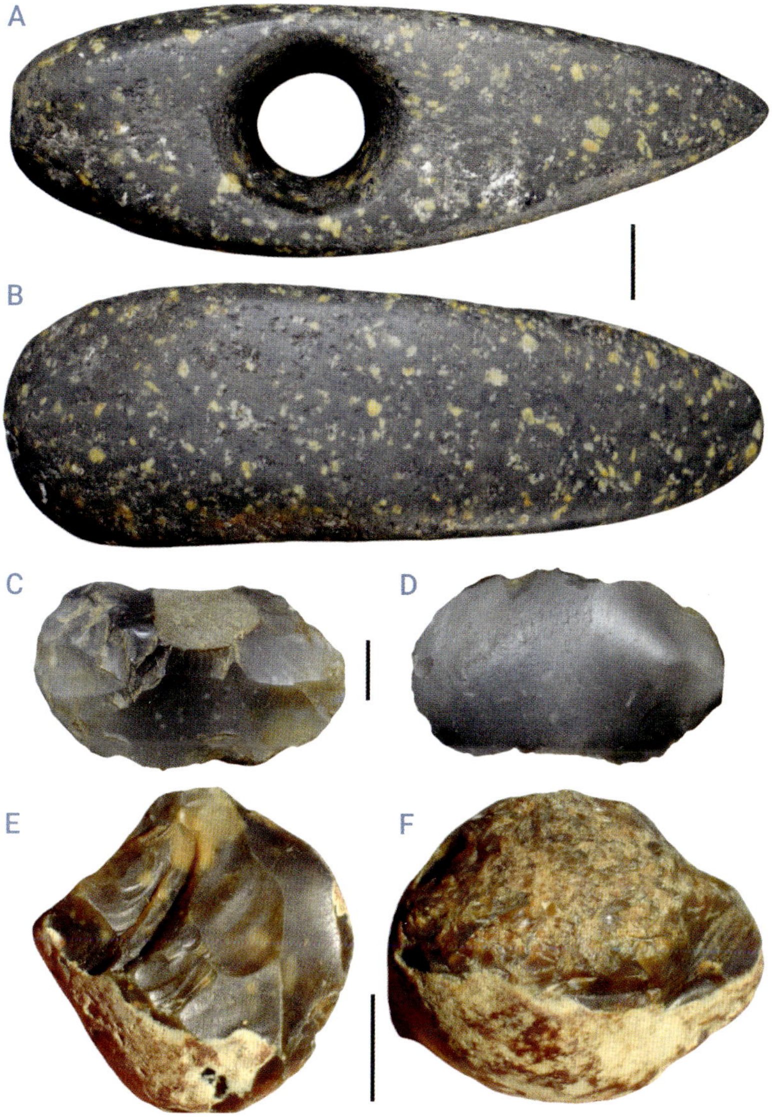

Figure 7.116

Neolithic and Bronze Age worked tools. A, B, Late Neolithic/ Early Bronze Age perforated Welsh stone ceremonial axe head. C, D, Neolithic/Early Bronze Age Scraper. E, F, Late Neolithic/Early Bronze Age multi-platform Core re-purposed as hammerstone. A, D, collection of Andrew Jacobs. E, F, collection of Theo Vickers. Scale equals 10 mm.

accessible. Aside from the odd use of a sharp flake or quickly made scraping tool flint tools probably ceased to be used commonly from 3,500 years ago.

Fire cracked flints are a somewhat common occurrence, and possible evidence of stone boiling. This is a process where ancient humans heated stones in a fire until they were red hot, then dropped them into a container (a hide lined container or a wooden trough)

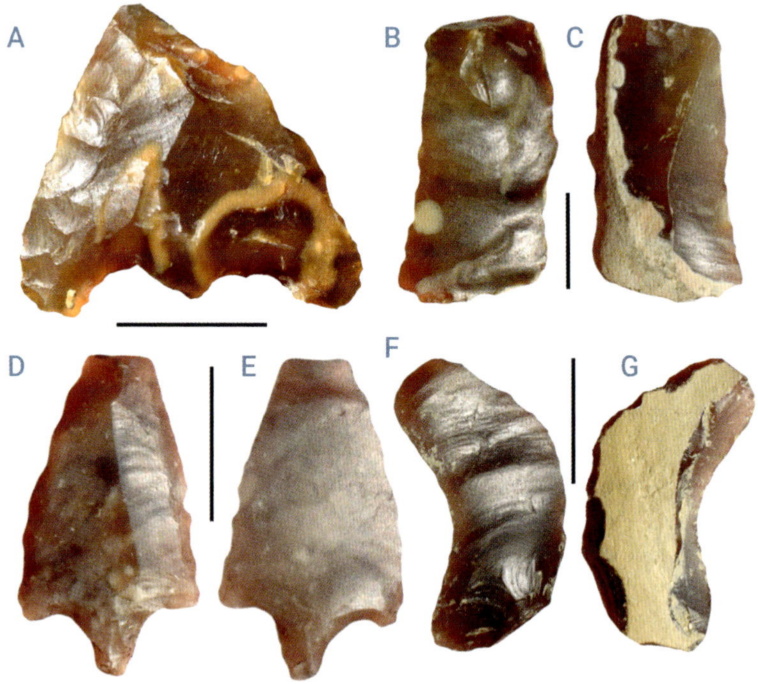

Figure 7.117

Neolithic and Bronze Age flint tools.

A, Late Neolithic/Early Bronze Age Barbed and Tanged Arrowhead. B, C, Late Neolithic/Early Bronze Age Backed Knife. D, E, Late Neolithic/Early Bronze Age Barbed and Tanged Arrowhead. F, G, Late Neolithic/Early Bronze Age Side Scraper. A, collection of Adie Butler. B-G, collection of Theo Vickers. Scale equals 10 mm.

of water to bring it to a boil. The heating process caused the flints to change their colour and texture, with results depending on the duration of heating, temperature they reached as well as how many times they were reused. Fire cracked stones can also be the result of intense and rapid fires (forest fires, heathland fires). It is suggested that a red colourings is linked to rapid intense fires, whereas a bleached colour is the result of controlled burning (*Fig. 7.118*).

The Longstone

Evidence for ancient human activity on the island not only comes in the form of flint tools but also the marks on the landscape

Figure 7.118

A, B, examples of pottery dating to the Beaker People dating between 25,00 – 1,200 BC, with ornamentation made using bone combs. C, fire cracked flint. A, B, collection of Theo Vickers. Scale equals 10 mm.

Figure 7.119

The Longstone standing stones at Mottistone.

our ancestors left behind. The most notable is the Longstone in Mottistone (*Fig. 7.119*), the island's only megalithic monument. The Longstone consists of two large iron cemented sandstones from the Lower Greensand Group, most likely carved from the rocks a few hundred metres north of the stones today. Today one remains standing, while the second stone has collapsed. These stones have been surrounded in myth and legend for hundreds of years. It was thought that the stones, which are out of place with the surrounding geology, were a result of a contest between the Devil and St Catherine to see who should control the Isle of Wight. Supposedly, the tallest pillar was thrown by St. Catherine, from the downs in the east that now bear her name. Satan's smaller stone fell short, thus losing the contest. Older versions of the story tell of giants placing the stones there, or giants fighting with the devil.

Despite these colourful local legends, the stones are the remains of an early Neolithic long barrow, up to 6,000 years old. Long barrows appear to represent communal tombs for early Neolithic farming communities and can be found across southern England at this time. In the early Neolithic, both stones would have been standing, marking an east-facing entrance to the monument. The barrow would have been 31m long, 9m wide and 2m high, and it is thought that 'sky burials' may have taken place at the site, with the deceased laid atop the barrow or a nearby platform, allowing for scavenging by wild animals and birds. Remains were then collected and placed within chambers, and perhaps occasionally retrieved and returned for rituals or ancestor worship. Despite this, excavations at the Longstone have been unfruitful in finding remains or significant artefacts except small flint tools.

The Longstone is also thought to have been a meeting point during the Anglo-Saxon period (around 410 – 1066 AD), The Old English for meeting place is 'moot' and it is thought Mottistone, is the corruption of 'moot stone'. The Longstone is now a scheduled and protected monument under the care of the National Trust, providing a rare glimpse into the culture of the island's first farmers.

Bronze Age round barrows

Round barrows, or tumuli are later burial monuments dating to the early Bronze Age, between 4000 - 3500 years ago, and consist of a mound of earth and stones raised over one or several graves. Across the island, round barrows play a significant role in our landscapes, often seen as prominent features along our chalk downs, and over 240 have been identified across the island. Round barrows likely served as elite burial sites, reserved for figures such as chieftains, warriors or religious leaders and their families, and may have also been territorial markers and ritual centres for communities. At times, later burials would also be inserted into pre-existing barrows, perhaps burying people in the barrow of their ancestors.

On the island three different types of barrow are present: bowl barrows (a simple mound of earth), disk barrows (a banked mound with an inner ditch and platform containing one or more burials)

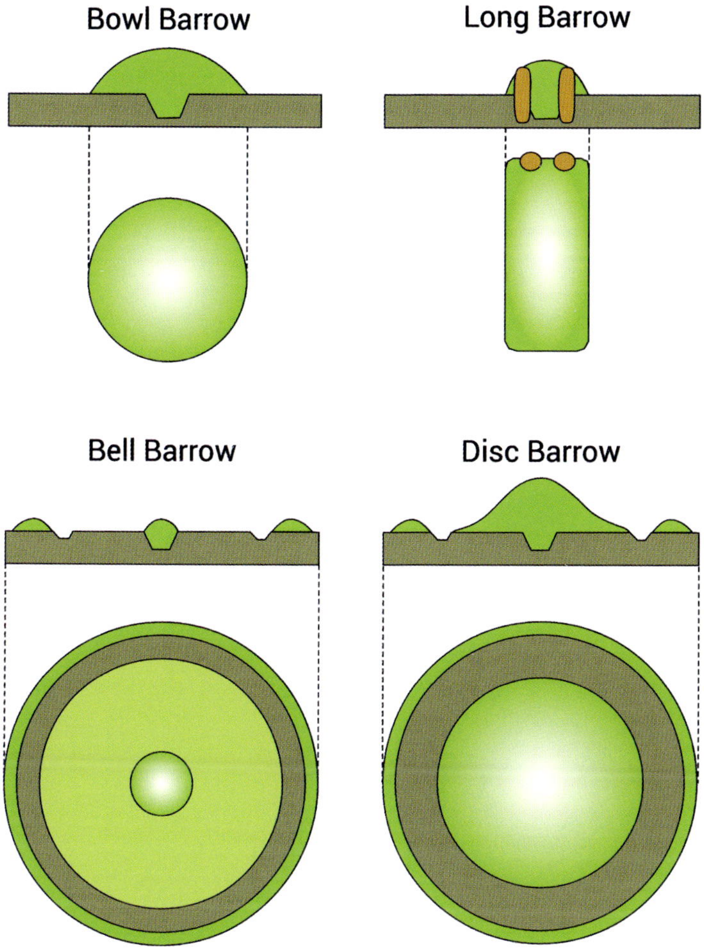

Figure 7.120

Diagram of different barrows found on the Isle of Wight

and a bell barrow (a large mound surrounded by a ditch and bank) (*Fig. 7.120*). The ditch is typically the source of the material to build the mound). Largely constructed on the island's chalk downland, when freshly built and in use these monuments would have appeared as striking white mounds and rings in their prominent positions on the landscape. Barrows eventually went out of use, but in the Anglo-Saxon and Viking periods, it wasn't uncommon for Bronze Age barrows to be re-purposed for burials, with early medieval individuals being placed into monuments already 2000 years old. Today many barrows also have large craters in the centre likely where medieval looters or 19th century archaeologists dug into them in search of artefacts, and many more have probably been completely lost to centuries of ploughing and farming, now reduced to subtle crop marks.

The most famous and accessible barrow group on the island is Five Barrows, situated at the eastern end of Brook Down. Despite its name, the site consists of an impressive cemetery of eight early Bronze Age barrows of varying types, size and probably date. Other accessible and visible barrows can be found at Headon Hill, Mottistone Down, Ashley Down, Brighstone Forest and Ventnor Down.

7.8. Imported fossils and other finds.

Not all fossils found on the island are from the island's geology. Some have been imported through sea defences construction, while others have been accidental through shipwrecks.

Sea defences.

On the western side of Ventnor beach are large blocks of Mendip Hill carboniferous limestone, around ~350 – 320 million years old. The fossils can be seen on the surface of the boulders, including crinoids, bryozoans, brachiopods, corals and gastropods (*Fig. 7.121*).

Great care must be taken when looking over the rocks forming sea defences as many are covered in very slippery algae. Do not try to extract any fossils, the limestone blocks are incredibly hard, and hammers simply bounce off the surface.

Ship ballast.

Ships prior to the 20th century contained a hard ballast, commonly made of stone and sand. This was placed low in a ship's hull when the cargo was being unloaded, keeping the boat at the right height in the water. Otherwise, it would pop out of the water, a bit like a cork, and become very unstable. However, after the introduction of steel hulled ships, and pumping systems, water is now used instead of hard ballast.

On the island, with some 2000 shipwrecks on our coastlines, it's very common to find pieces of this ballast, usually granites and

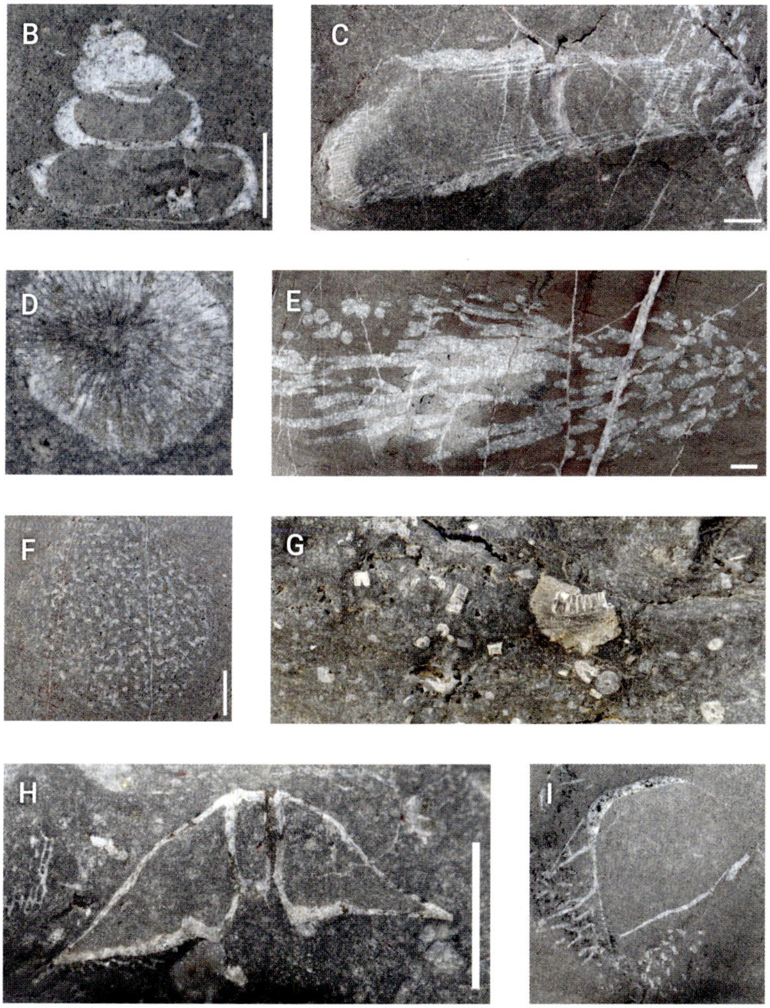

Figure 7.121

A, the sea defences at Ventnor. B, gastropod. C, lengthways cross section of a rugose coral. D, cross section of a rugose coral. E, rugose coral. F, ?sponge coral. G, disarticulated crinoid stems. H, cross section of a sphiriferid brachiopod. I, cross section of a brachiopod encrusted by a tabulate coral. Scale equals 10 mm.

Figure 7.122

Other finds.

A, tarmac. B, brick. C, concrete water pipe. D, rock from a gabion basket (sea defence). E, H, different types of sandstone. F, encrusted metal. G, J, examples of ship ballast. I, iron and calcite nodule. Scale equals 10 mm.

vein quartz amongst other igneous and metamorphic rocks, with the main indication being they're not a local rock type (*Fig. 7.122*). Pebbles of Carboniferous limestone containing crinoids can also be found.

Other finds.

Our beaches also see many finds which are not fossilised remains but do appear to be interesting to the untrained eye, usually the remains of manmade structures or purely geological. The most common occurrences are pieces of concrete, tarmac, roof tile, bricks, pipes, sandstone, and pieces of encrusted rusted metal (*Fig. 7.122*).

Figure 7.123

A, calcite crystals from Whale Chine. B, cross section of 'Beef' showing the cone in cone structure. C, calcite crystals from the Bembridge Limestone from Fort Victoria. D, 'Isle of Wight Diamond', selenite crystal from Atherfield. E, dorsal view of 'Beef' showing the circular pattern of the cones. F, cone in Cone calcite. G, Selenite crystal from Bouldnor. H, I, vivianite crystals from Whale Chine. A, H, I, collection of Jack Wonfor. C, collection of Theo Vickers. Scale equals 10 mm.

7.9. Minerals

Minerals are naturally occurring solid inorganic substances with a defined chemical composition and a regular crystal structure. Often seen as polished pebbles in fossil shops, or large and impressive crystals such as quartz. Minerals make up all rocks and are often involved in the fossilisation process but can be found as larger crystals that have formed within the rocks on the island. Within our rocks, several different minerals can be found and are discussed below.

Selenite – Wight Diamonds

If you are walking along the beach between Shepherd's Chine and Atherfield point on a sunny day, you may see some bright sparkling coming from the steep grey cliff. These are Isle of Wight diamonds signifying their presence.

These large crystals have long been called 'Isle of Wight diamonds' due to their shape, but are in fact crystals of selenite, which is a variety of gypsum (calcium sulphate). In contrast to diamonds, selenite is one of the softest minerals known and can be scratched with your fingernail (*Fig. 7.123*).

Selenite forms when iron pyrite that is scattered as fine grains throughout the Atherfield Clay oxidises. The FeS2 of the pyrite forms dilute sulphuric acid, which in turn reacts with calcium carbonate also in the clay and produces calcium sulphate. When the calcium sulphate crystalises it forms selenite. Gypsum minerals are commonly found in alkaline lake muds, clays, and evaporated seas. Gypsum does dissolve in water, albeit very slowly, so make sure you keep them dry and away from water after you have cleaned the clay off.

Calcite

Calcite is a carbonate mineral (CaCO3) and is a very abundant. It's an evaporite mineral, meaning it forms when waters that contain concentrated dissolved salts, evaporate, causing the minerals to precipitate out and form the minerals we see today. On the island, calcite can be found infilling chambers of ammonites from the Lower Greensand, within the cavities of dinosaur bones and as geodes in limestone blocks on the north coast of the island (*Fig. 7.123*).

'Beef'– cone-in-cone structures.

The Vectis Formation occasionally yields 'Beef' or 'cone-in-cone' structures (*Fig. 7.123*).

Cone-in-cone or 'beef' refers to conical and occasionally fibrous looking structures, that are arranged vertically within a thin bed. The formation of these structures is still debated, but the 'displacive crystal growth mechanism' is widely accepted. This is where crystals start to form in partially consolidated sediment. As the crystal structures grow, they begin to take up more and more space, causing pressure to build within the sediment. The pressure varies, causing the conical structure to form.

The occurrence of the structures was initially described over 200 years ago by famous geologists such as Thomas Webster, William Buckland, and Henry De la Beche. The term 'Beef' was coined by quarry workers in Purbeck, Southern England, after it's fibrous similarity to animal muscle. Some of the most famous examples of 'Beef' come from the Lyme Regis area, with a layer within the Lias called Shales-with-Beef Member within the Charmouth Mudstone Formation.

Vivianite

Vivianite is a hydrated iron phosphate mineral, (Fe2+Fe2+2(PO4)2·8H2O). When vivianite is fresh, it appears colourless. But after exposure to oxygen, the Fe2+ converts to Fe3+ causing the mineral to darken and turn dark blue to blue green. After being left in the light, it may turn almost black. Vivianite is a secondary mineral, meaning they're formed as a byproduct of the breakdown of the primary mineral (those present when the rock forms) or a chemical/ physical reaction within the rock.

Vivianite crystals are found on rare occasions on ammonites from the Lower Greensands, as well as within the cracker nodules where they occur as deep blue or even black prismatic or flattened crystals (*Fig. 7.123*).

Pyrite

Pyrite is an iron disulphide mineral, FeS2. It's metallic lustre and brassy hue gives it a superficial resemblance to gold, hence the common name 'fool's gold'. The name pyrite derives from the Greek pyritēs lithos, meaning stone or mineral which strikes fire, as when it's struck against other materials, it creates sparks. Pyrite occurs naturally, often replacing fossils leaving a pyrite infill or mould, but it also occurs very commonly as a variety of different crystal forms on the island (*Fig. 7.124*).

Figure 7.124

A, quartz crystals in a flint pebble. B, chalcedony crystals in a flint pebble. C, polished agate pebble from Bembridge. D, polished agate from Binstead set into a crab necklace. E, round pyrite ball from the Vectis Formation. F, pyrite from the Vectis Formation infilling trace fossils. G, internal structure of marcasite from the chalk at Yaverland. H, external surface of marcasite from the chalk at Yaverland. I, cubic crystal growth of pyrite from the Wessex Formation. C, D, collection of Rick Lilley and Charlotte Willis. Scale equals 10 mm.

Rainbow pyrite concretions can be found at certain horizons at Bouldnor, Hamstead in the Paleogene formations and within the Vectis Formation. The cracks within a concretion are lined with iridescent pyrite microcrystals and are visible when the concretions are broken open.

Pyrite spheres within the Vectis Formation are often found weathered on the foreshore and usually a metallic green colour. These form when the pyrite crystalises as bands within the sediment.

The pyrite from the Vectis Formation can also take on very 'bobbly' and irregular shapes and is often mistaken for coprolites. Coprolites will always be dark brown – black in colour and will weigh substantially less than an equivalently sized pyrite pebble.

The Wessex Formation plant debris beds often yield large clusters of pyrite crystals, sometimes attached to the fossil wood. These are a mixture of cubic, pyritohedron and octahedron crystals.

The Vectis Formation yields brown concretions with large cubic pyrite crystals.

Marcasite

Although having the same composition as pyrite (FeS2), marcasite is physically and crystallographically distinct from pyrite, caused by differences in the arrangement of sulphur atoms around the iron atoms in the crystal lattice. This difference at an atomic level causes it to crystallise in the orthorhombic system rather than the cubic system. This makes marcasite lighter and more brittle than pyrite. It is also worth noting that marcasite decays more readily than pyrite, so it should be kept dry at all times.

Marcasite most commonly occurs on the island in the chalk as dark brown nodules in the lower part of the chalk. These are often mistaken for meteorites due to the irregular surface and weight. However, once cracked open, the individual crystals radiating from a central point with a brassy metallic shine becomes obvious (*Fig. 7.124*).

Agates

Agates are a banded variety of chalcedony. The banding is caused by alternating layers of crystalline quartz. They often show a wide range of colours and banding patterns, which are more obvious when the pebbles are wet or polished (*Fig. 7.124*). Although agates can be found on the islands' beaches, they do not originate here. It's possible they were part of ballast from sunken ships or brought here in the Ice Age river gravels, perhaps from the West Country.

Where to see Isle of Wight fossils

Dinosaur Isle Museum, Sandown currently houses the registered collection on the island, with many holotypes on permanent display, including the dinosaurs *Neovenator* and *Eotyrannus*.

Dinosaur Expeditions C.I.C, locally known as 'the Dinosaur Farm', is located on the Military Road, Brighstone. This is a great little museum, with excellent examples of the island's fossils on display.

Natural History Museum, London, houses many fossils from the island in its collections, and some are on public display, such as the skeleton of the dinosaur *Mantellisaurus* currently on display in the main gallery.

Smaller attractions, that aren't dedicated to palaeontology sometimes hold hidden fossils within their displays. Examples include the Swiss Cottage at Osborne House (English Heritage), which holds a variety of fossils—including some from the Isle of Wight—and Newtown Old Town Hall (National Trust), which displays bones from Ice Age mammals found locally. The heritage centre in Shanklin Chine also features dinosaur bones and ammonites.

Fossils from the island have found themselves in many museums across the UK, but also around the world, so it's always worth keeping an eye out for Isle of Wight on the museum labels. One example is the Saurier museum, Aathal, Switzerland. This is an amazing museum full of fossils, focusing on dinosaurs, with a whole room dedicated to UK dinosaurs. Definitely worth a visit for any fossil or dinosaur enthusiast!

Acknowledgements

Fossil Collectors. For their dedication to the subject, the hours spent on our shores rescuing fossils in all weathers, all year round, and donating their finds to museums. Their time and energy have led to countless advancements in our scientific understanding of the island's palaeontological past.

Researchers. For making many advances in our understanding of the geology and palaeontology of the Isle of Wight available in the scientific literature.

Museums. The dedicated staff over the years did much to conserve and preserve the island's palaeontological heritage, and special mention must be made of the curators and staff of the new Dinosaur Isle Museum, as well as all the curators since the formation of the museum in 1818.

Personal acknowledgements.

I would firstly like to thank my family, Andrew, Clare, Aaron, Derek, Eve, Moira and all my friends for all their ongoing support, high tolerance of my fossil ramblings and encouragement to pursue a palaeontological career.

A massive thank you to my mum for her support over the years, with the 'just go for it and see' attitude, allowing me to travel across the world to follow my passion. But most importantly for tolerating all the rocks around the house (and in the garden), clay covered surfaces and numerous dinosaur bones on the kitchen table. I expect you were hoping it was a 'phase' I'd grow out of.

To dad, thank you for the time spent outdoors, finding fossils, walking the beaches, and fishing, creating the best childhood anyone could ask for. And ultimately led me down the path of scientific curiosity to where I am today and I'm so grateful I can share it with you.

I am greatly indebted to Andy Gale, for his friendship, generosity, encouragement, support, and advice, all whilst keeping me sane, housed and well fed. As well as kindly proofreading the text, fact checking, assisting with figures and photography, and providing answers to all my questions and queries without hesitation.

A huge thankyou to David Martill, for his unfaltering belief in me. His generosity in time, support and encouragement over the years has allowed me to grow into a confident researcher and writer. And probably a bigger thank you for listening (and tolerating) all my grumbles and keeping me grounded. As well spending countless tedious hours proofreading the text.

To Oliver and Izzy Mattsson of Dinosaur Expeditions who, back in 2015, took a chance and let me lead fossil walks as a lowly undergraduate. Since then, both have become close friends and provide an immense amount of support and encouragement in everything I do, which I am incredibly grateful for.

I would also like to thank the following for kindly providing specimens to include and fossil identification: Adie Butler, Claire Hampstead, Charlotte Willis, Dick Downes, Glenn Martin, Huxley Hunt, Jack Wonfor, Jacque Bainbridge, Joe Thompson, Kai Baliey, Lucy Smith, Luke Ferguson, Maria Yule, Mick Green, Nathan Skene, Oliver Mattsson (Dinosaur Expeditions C.I.C), Rick Lilley, Roy Smith, Sam Caethoven, Scott Cade and Theo Vickers.

And for advice, discussions and support over the years, I'd like to thank Alex Peaker, Andy Yule, Andy Ashdown, Ben Thomas, Darren Naish, David Loydell, Dick Downes, Gary Blackwell, Gary Hoare, Jerry Hooker, Jack Wonfor, Jeremy Lockwood, Jon and Lou Moody, Josie Pallister, Kai Bailey, Keith Simmonds, Mark Penn, Martin Munt, Martyn Hornett, Mick Green, Neil Gostling, Nick Longrich, Nizar Ibrahim, Oliver Mattsson, Remmert Schouten, Roy Smith, Steve Etches (and the staff of the Etches Collection), Steve Sweetman, Steve Burbridge, Susanna Hoe, Shaun Smith, Silas Shaul, Theo Vickers and Tony Butcher. A big thankyou to Martin Munt for checking my mollusc identifications, David Ward for shark related identifications, and James Brown (National Trust) for help with archaeological questions.

To Anne and Sophie for reading through the book providing a non-academic perspective and helpful edits. A massive thankyou to Neil Gostling for reading through and providing an excellent review. A special thanks to Mick Green for the hours spent chatting about the Wessex Formation and providing countless cups of coffee in the process.

I must say a special thank you to those who are no longer with us and are greatly missed by many.

Firstly Nick Chase, who always showed great enthusiasm to impart his knowledge, and never hesitated to walk the beaches with me. His generosity with his time and the impressive 6th sense for dinosaur bones will always remain an important part of my early career and entry into the world of dinosaur hunters on the island.

And Paul Newtown, his infectious passion for fossils opened my eyes to the island's Lower Greensand. His encouragement and advice led me to discover hidden gems within the south coast of the island and showed me there's more than dinosaurs on the island!

I would like to thank Gary Meek, Andy Ashdown, Jenny Courtney and the IOW National Trust team for all their support and kindness, allowing us to work together to really highlight the geology and palaeontology of the island's beaches, as well as making the summers at Compton fun and productive. And not forgetting Gary and Phil in the ice cream vans!

A massive thank you to Kirby Siber for sponsoring the book's publication.

I am indebted to my friends and business partners, Jack Wonfor and Theo Vickers. Together we started a crazy journey starting up our own fossil business at a kitchen table one evening, not even imagining Wight Coast Fossils would be where it is today. And our guides Nelle Findley, Joe Thompson and Huxley Hunt for all their help and for joining the adventure!

An enormous thankyou to Peter Harrigan, Simon Harrop, Charlotte Beadle, Richard Bull, Gurjeet Mathadu and Sherif Dhaimish at Medina Publishing for all their support, enthusiasm and hard work to publish the book!

I am incredibly lucky to say walking the beautiful island's beaches is my job, which wouldn't be possible without supporters and followers of Wight Coast Fossils. So, a massive thank you to everyone who supports us via social media, books onto our walks, pre ordered or donated on our crowdfunder, or buys this book! Your desire and passion to learn more about our island's prehistoric past led to the writing of this book.

Further Reading

Austen, P. A., & Batten, D. J. (2018). English Wealden fossils: An update. *Proceedings of the Geologists' Association, 129*(2), 171–201.

Barker, C. T., Hone, D. W., Naish, D., Cau, A., Lockwood, J. A., Foster, B., Clarkin, C. E., Schneider, P., & Gostling, N. J. (2021). New spinosaurids from the Wessex Formation (Early Cretaceous, UK) and the European origins of Spinosauridae. *Scientific Reports, 11*(1), 19340.

Barker, C. T., Lockwood, J. A., Naish, D., Brown, S., Hart, A., Tulloch, E., & Gostling, N. J. (2022). A European giant: A large spinosaurid (Dinosauria: Theropoda) from the Vectis Formation (Wealden Group, Early Cretaceous), UK. *PeerJ, 10*, e13543.

Batchelor, T. J., & Duffin, C. J. (2020). First description of sharks' teeth from the Ferruginous Sands Formation (Aptian, Early Cretaceous) of the Isle of Wight. *Proceedings of the Geologists' Association, 131*(3–4), 353–359.

Batten, D. J. (Ed.). (2011). *English Wealden fossils.* Palaeontological Association.

Beckles, S. H. (1851). On supposed casts of footprints in the Wealden. *Quarterly Journal of the Geological Society, 7*, 117.

Beckles, S. H. (1854). On the Ornithoidichnites of the Wealden. *Quarterly Journal of the Geological Society, 10*, 456–464.

Beckles, S. H. (1862). On some natural casts of footprints from the Wealden of the Isle of Wight and Swanage. *Quarterly Journal of the Geological Society of London, 18*, 443–447.

Clabby, S. M. (2001). *Dino Wight: The dinosaurs of the Isle of Wight.*

Cohen, K. M., Harper, D. A. T., Gibbard, P. L., & Car, N. (2023). ICS International Chronostratigraphic Chart 2023/09.

Colenutt, G.W. (1888). On a portion of the Osborne Beds of the Isle of Wight, and on some remarkable organic remains recently discovered therein. *Geological Magazine*, 25, 358–362.

Coram, R. A., Radley, J. D., & Martill, D. M. (2017). A Cretaceous calamity? The Hypsilophodon bed of the Isle of Wight, southern England. *Geology Today, 33*(2), 66–70.

Delair, J. B. (1983). Cretaceous dinosaur footprints from the Isle of Wight: A brief history. *Proceedings of the Isle of Wight Natural History and Archaeological Society VII*, 600–615.

Drake, E. T. (2007). The geological observations of Robert Hooke (1635–1703) on the Isle of Wight. *Geological Society, London, Special Publications, 287*, 19–30.

Francis, J. E., & Harland, B. M. (2006). Termite borings in Early Cretaceous fossil wood, Isle of Wight, UK. *Cretaceous Research, 27*(6), 773–777.

Gale, A. S., Huggett, J. M., & Gill, M. (1996). The stratigraphy and petrography of the Gault Clay Formation (Albian, Cretaceous) at Redcliff, Isle of Wight. *Proceedings of the Geologists' Association, 107*(4), 287–298.

Gale, A. (2019). *The Isle of Wight: Geologists' Association Guide No. 6.* The Geologists' Association.

Gale, A. S. (2021). The stratigraphy of the upper Campanian Chalk of the southern English coast (Isle of Wight, Dorset), United Kingdom. *Cretaceous Research, 124*, 104775.

Hayes, P. A., & Collinson, M. E. (2013). The flora of the Insect Limestone (latest Eocene) from the Isle of Wight, southern England. *Earth and Environmental Science Transactions of the Royal Society of Edinburgh, 104*(3–4), 245–261.

Hopson, P. (2011). The geological history of the Isle of Wight: An overview — a synthetic overview helpful for context and stratigraphic correlations.

Hopson, P. M., & Farrant, A. R. (2015). Geology of the Isle of Wight – A brief explanation of the geological sheet. *Sheet Explanation of the British Geological Survey.* British Geological Survey, Keyworth.

Hooker, J. J., Evans, S. E., & Davis, P. G. (2019). Vertebrate remains from the Insect Limestone (latest Eocene), Isle of Wight, UK. *Earth and Environmental Science Transactions of the Royal Society of Edinburgh, 110*(3–4), 281–287.

Hooley, R. (1907). A brief sketch of the Wealden beds of the Isle of Wight, and the history they reveal. *Papers and Proceedings of the Hampshire Field Club, 6*, 90–105.

Hooley, R. W. (1913). On the skeleton of *Ornithodesmus latidens*; an ornithosaur from the Wealden Shales of Atherfield (Isle of Wight). *Quarterly Journal of the Geological Society, 69*(1–4), 372–422.

Hooley, R. W. (1925). On the skeleton of *Iguanodon atherfieldensis* sp. nov., from the Wealden Shales of Atherfield (Isle of Wight). *Quarterly Journal of the Geological Society, 81*(1–4), 1–61.

Huggett, J. M., Gale, A. S., & Hazell, C. (2018). Early diagenesis of the lower Vectis Formation, Wealden Group (Lower Cretaceous, Barremian), Sandown, Isle of Wight, southern England. *Proceedings of the Geologists' Association, 129*(5), 611–624.

Hulke, J. W. (1871). Note on a large reptilian skull from Brooke, Isle of Wight, probably dinosaurian, and referable to the genus *Iguanodon. Quarterly Journal of the Geological Society, 27*(1–2), 199–206.

Hulke, J. W. (1873). Contribution to the anatomy of *Hypsilophodon foxii*: An account of some recently acquired remains. *Quarterly Journal of the Geological Society, 29*(1–2), 522–532.

Hulke, J. W. (1887). VII. Supplemental note on *Polacanthus foxii*, describing the dorsal shield and some parts of the endoskeleton, imperfectly known in 1881. *Philosophical Transactions of the Royal Society of London. B, 178*, 169–172.

Hutt, S., Martill, D. M., & Barker, M. J. (1996). The first European allosauroid dinosaur (Lower Cretaceous, Wealden Group, England). *Neues Jahrbuch für Geologie und Paläontologie - Monatshefte, 1996*(10), 635–644.

Hutt, S., Naish, D., Martill, D. M., Barker, M. J., & Newbery, P. (2001). A preliminary account of a new tyrannosauroid theropod from the Wessex Formation (Early Cretaceous) of southern England. *Cretaceous Research, 22*(2), 227–242.

Huxley, T. H. (1870). On *Hypsilophodon foxii*, a new dinosaurian from the Wealden of the Isle of Wight. *Quarterly Journal of the Geological Society, 26*(1–2), 3–12.

Joyce, W. G., Chapman, S. D., Moody, R. T., & Walker, C. A. (2011). The skull of the solemydid turtle *Helochelydra nopcsai* from the Early Cretaceous of the Isle of Wight (UK) and a review of Solemydidae. *Special Papers in Palaeontology, 86*, 75–97.

Lockwood, J. A., Lockley, M. G., & Pond, S. (2014). A review of footprints from the Wessex Formation (Wealden Group, Lower Cretaceous) at Hanover Point, the Isle of Wight, southern England. *Biological Journal of the Linnean Society, 113*(3), 707–720.

Lockwood, J. (2016). Ichnological evidence for large predatory dinosaurs in the Wessex Formation (Wealden Group, Early Cretaceous) of the Isle of Wight. *Proceedings of the Isle of Wight Natural History and Archaeological Society, 30*, 103–110.

Lockwood, J. A., Martill, D. M., & Maidment, S. C. (2021). A new hadrosauriform dinosaur from the Wessex Formation, Wealden Group (Early Cretaceous), of the Isle of Wight, southern England. *Journal of Systematic Palaeontology, 19*(12), 847–888.

Lockwood, J. A., Martill, D. M., & Maidment, S. C. (2024). *Comptonatus chasei*, a new iguanodontian dinosaur from the Lower Cretaceous Wessex Formation of the Isle of Wight, southern England. *Journal of Systematic Palaeontology, 22*(1), 2346573.

Lockwood, J. A., Martill, D. M., & Maidment, S. C. (2025). The origins of neural spine elongation in iguanodontian dinosaurs and the osteology of a new sail-back styracosternan (Dinosauria, Ornithischia) from the Lower Cretaceous Wealden Group of England. *Papers in Palaeontology, 11*(4), e70034.

Lomax, D. R., & Tamura, N. (2014). *Dinosaurs of the British Isles* (pp. 414). Manchester: Siri Scientific Press.

Longrich, N. R., Martill, D. M., & Jacobs, M. L. (2022). A new dromaeosaurid dinosaur from the Wessex Formation (Lower Cretaceous, Barremian) of the Isle of Wight, and implications for European palaeobiogeography. *Cretaceous Research, 134*, 105123.

Longrich, N. R., Martill, D. M., Munt, M., Green, M., Penn, M., & Smith, S. (2024). *Vectidromeus insularis*, a new hypsilophodontid dinosaur from the Lower Cretaceous Wessex Formation of the Isle of Wight, England. *Cretaceous Research, 154*, 105707.

Mannion, P. D., Upchurch, P., & Hutt, S. (2011). New rebbachisaurid (Dinosauria: Sauropoda) material from the Wessex Formation (Barremian, Early Cretaceous), Isle of Wight, United Kingdom. *Cretaceous Research, 32*(6), 774–780.

Mantell, G. A. (1846). Notes on the Wealden strata of the Isle of Wight, with an account of the bones of *Iguanodons* and other reptiles, discovered at Brook Point and Sandown Bay. *Quarterly Journal of the Geological Society of London, 2*, 91–96.

Mantell, G. A. (1847). *Geological excursions around the Isle of Wight, and along the adjacent coast of Dorsetshire: Illustrative of the most interesting geological phenomena and organic remains* (3rd ed.). London: Bohn.

Marsden, M. K., Gunn, J., Maidment, S. C., Nichols, G., Wheeley, J. R., Russell, C. E., Boomer, I., Stukins, S., & Butler, R. J. (2025). Palaeoenvironment and taphonomy of the *Hypsilophodon* Bed, Lower Cretaceous Wessex Formation, Isle of Wight. *Journal of the Geological Society, 182*(1), jgs2024-046.

Martill, D. M., Frey, E., Green, M., & Green, M. E. (1996). Giant pterosaurs from the Lower Cretaceous of the Isle of Wight, UK. *Neues Jahrbuch für Geologie und Paläontologie Monatshefte*, 672–683.

Martill, D. M., & Naish, D. (2001). *Dinosaurs of the Isle of Wight.* Palaeontological Association Field Guides to Fossils Series, No. 10.

Martill, D. M. (2015). First occurrence of the pterosaur *Coloborhynchus* (Pterosauria, Ornithocheiridae) from the Wessex Formation (Lower Cretaceous) of the Isle of Wight, England. *Proceedings of the Geologists' Association, 126*(3), 377–380.

Morton, A. (2001). A collection of Eocene and Oligocene fossils.

Munt, M. C. (2016). Geoheritage case study: The Isle of Wight, England. In T. A. Hose (Ed.), *Geoheritage and Geotourism: A European perspective* (pp. 195–204). Boydell Press.

Naish, D., Martill, D. M., Cooper, D., & Stevens, K. A. (2004). Europe's largest dinosaur? A giant brachiosaurid cervical vertebra from the Wessex Formation (Early Cretaceous) of southern England. *Cretaceous Research, 25*(6), 787–795.

Norman, M. W. (1887). *A popular guide to the geology of the Isle of Wight with a note on its relation to that of the Isle of Purbeck.* Knight's Library, Ventnor.

Osbourne White, H. J. (1921). *A short account of the geology of the Isle of Wight.* HM Stationery Office.

Owen, R. (1842). Report on British fossil reptiles, part 2. *Report of the British Association for the Advancement of Science, 11,* 60.

Owen, R. (1853). *Monograph on the fossil Reptilia of the Wealden and Purbeck formations* (Vol. 7). Palaeontographical Society.

Pond, S., Lockley, M. G., Lockwood, J. A. F., Breithaupt, B., & Matthews, N. (2014). Tracking dinosaurs on the Isle of Wight: A review of tracks, sites, and current research. *Biological Journal of the Linnean Society, 113,* 737–757.

Pond, S. (2017). The most complete ankylosaur skull ever found in the Wessex Sub-basin (Lower Cretaceous) of the Isle of Wight. *PeerJ PrePrints.*

Pond, S., Strachan, S. J., Raven, T. J., Simpson, M. I., Morgan, K., & Maidment, S. C. (2023). *Vectipelta barretti,* a new ankylosaurian dinosaur from the Lower Cretaceous Wessex Formation of the Isle of Wight, UK. *Journal of Systematic Palaeontology, 21*(1), 2210577.

Price, T. (2014). New dinosaur footprints exposed in rocks of the Wessex Formation, Lower Cretaceous, at Sandown, Isle of Wight, southern England. *Biological Journal of the Linnean Society, 113*(3), 758–769.

Quayle, W. J., & Collins, J. S. H. (2012). A review of the decapod crustaceans from the Tertiary of the Isle of Wight, Hampshire, UK, with description of three new species. *Bulletin of the Mizunami Fossil Museum, 38,* 33–51.

Quayle, W. J. (2020). Nominated neotype specimen for a decapod crustacean, Neptunus vectensis Carter, 1898 from the Oligocene of the Isle of Wight, Hampshire, UK. *Bulletin of the Mizunami Fossil Museum, 47*, 45–48.

Radley, J. D., Barker, M. J., & Harding, I. C. (1998). Palaeoenvironment and taphonomy of dinosaur tracks in the Vectis Formation (Lower Cretaceous) of the Wessex Sub-basin, southern England. *Cretaceous Research, 19*(3-4), 471–487.

Radley, J. D., Gale, A. S., & Barker, M. J. (1998). Derived Jurassic fossils from the Vectis Formation (Lower Cretaceous) of the Isle of Wight, southern England. *Proceedings of the Geologists' Association, 109*(2), 81–91.

Robinson, S. A., & Hesselbo, S. P. (2004). Fossil-wood carbon-isotope stratigraphy of the non-marine Wealden Group (Lower Cretaceous, southern England). *Journal of the Geological Society, 161*(1), 133–145.

Selden, P. A. (2002). First British Mesozoic spider, from Cretaceous amber of the Isle of Wight, southern England. *Palaeontology, 45*(5), 973–983.

Smith, A. B., & Batten, D. J. (Eds.). (2002). *Fossils of the Chalk*. Palaeontological Association.

Steel, L., Martill, D. M., Unwin, D. M., & Winch, J. D. (2005). A new pterodactyloid pterosaur from the Wessex Formation (Lower Cretaceous) of the Isle of Wight, England. *Cretaceous Research, 26*(4), 686–698.

Sweetman, S. C., & Insole, A. N. (2010). The plant debris beds of the Early Cretaceous (Barremian) Wessex Formation of the Isle of Wight, southern England: Their genesis and palaeontological significance. *Palaeogeography, Palaeoclimatology, Palaeoecology, 292*(3-4), 409–424.

Sweetman, S. C., & Martill, D. M. (2010). Pterosaurs of the Wessex Formation (Early Cretaceous, Barremian) of the Isle of Wight, southern England: A review with new data. *Journal of Iberian Geology, 36*(2), 225–242.

Sweetman, S. C., Goedert, J., & Martill, D. M. (2014). A preliminary account of the fishes of the Lower Cretaceous Wessex Formation (Wealden Group, Barremian) of the Isle of Wight, southern England. Biological Journal of the Linnean Society, 113(3), 872–896.

Torrens, H. S. (2014). The Isle of Wight and its crucial role in the 'invention' of dinosaurs. B*iological Journal of the Linnean Society, 113*(3), 664–676.

Wright, V. P., Taylor, K. G., & Beck, V. H. (2000). The paleohydrology of Lower Cretaceous seasonal wetlands, Isle of Wight, southern England. *Journal of Sedimentary Research, 70*(3), 619–632.

Young, J. R., Gale, A. S., Knight, R. I., & Smith, A. B. (Eds.). (2010). *Fossils of the Gault Clay* (Field Guides to Fossils No. 12). London: The Palaeontological Association.

Young, M. T., Steel, L., Foffa, D., Price, T., Naish, D., & Tennant, J. P. (2014). Marine tethysuchian crocodyliform from the Aptian-Albian (Lower Cretaceous) of the Isle of Wight, UK. *Biological Journal of the Linnean Society, 113*(3), 854–871.

Glossary

A

Aggradation. A process where the land elevation increases due to the accumulation of sediment being deposited, usually filling low lying areas or depressions.

Albian. A geological stage in the Lower Cretaceous extending from 113 – 100.5 million years before present.

Alligator. A large semiaquatic reptile, like a crocodile but with a broad and short head. Also known for saying 'see ya later'.

Anthracothere. The Anthracotheriidae. A family of extinct artiodactyls closely related and possibly ancestral to the hippopotami.

Anticline. An upward fold of stratified rock, creating a convex 'n' shape.

Artiodactyls. A group of hoofed mammals with an even number of toes on each foot.

Amber. Hard, translucent fossilised resin originating from coniferous trees.

Ankylosaur. Quadrupedal herbivorous dinosaur with bony armour and heavily built.

Aptian. A geological stage in the Lower Cretaceous extending from 121.4 – 113 million years before present.

Aragonite. A calcium carbonate mineral.

B

Barrow. Accumulation of earth or stones over a grave. Also known as burial mounds or tumuli.

Barremian. A geological stage in the Lower Cretaceous extending from 125.7 — 121.4 million years before present.

Basal. The bottom edge or surface of a bed/object or if referred to an animal, e.g., 'basal horse', meaning a very early form in that group's evolutionary history.

Basement. The rock surface beneath all sedimentary rocks.

Basin. A depression or dip in the earth's surface. This can also be the portion of land drained by a river and its tributaries.

Bedding surface. The surface that separates each layer from those above or below, usually recording a change in deposition.

Benthic. Refers to any animal, habitat or feature associated with living/occurring on the bottom of a body of water (lake, ocean).

Biogenic. Produced or brought about by living organisms.

Biological process. Chemical reactions or transformations that contribute to the overall functioning of the organism.

Bioturbation. Disturbance of sedimentary deposits by living organisms, such as burrowing or footprints.

Bivalve. An aquatic mollusc which has a compressed body enclosed within a hinged shell, such as oysters, mussels, and scallops.

Bony fish. A fish within the group Osteichthyes, that is distinguished by having skeleton of bones, scales, paired fins, one pair of gill openings, jaws and paired nostrils. This encompasses the vast majority of fish species.

Brackish. Water that is slightly salty, commonly found in river estuaries.

Brachiopod. A marine invertebrate of the phylum Brachiopoda, which comprises the lamp shells.

C

Calcareous. A sedimentary rock which is made from or contains a high proportion of calcium carbonate.

Calcite. A white or colourless mineral consisting of calcium carbonate. It is a major constituent of sedimentary rocks such as limestone and is deposited in caves to form stalactites and stalagmites.

Campanian. A geological stage in the Upper Cretaceous, extending from 83.6 – 72.1 million years before present.

Cartilaginous fish. A fish, within the group Chondrichthyes, that have a skeleton made from cartilage rather than bone (sharks and rays).

Caulkhead. Someone born and bred on the Isle of Wight, with at least 3 generations island heritage on both sides.

Cenomanian. A geological stage in the Upper Cretaceous, extending from 100.5 – 93.9 million years before present.

Cenozoic. The current geological era, representing the last 66 million years of Earth's history.

Chalk. A sedimentary limestone composed of calcite, formed in deep sea conditions from the accumulation of calcareous plankton.

Chemical weathering. When rainwater contains dissolved CO2, making it slightly acidic. This then causes reactions to occur when it comes into contact with minerals in the rock, altering the rock structure.

Chine. A local term for a V shaped cutting in the cliff, made by a stream eroding rapidly though the soft sands and clays.

Chitin. A natural polymer which is found in the exoskeleton of insects and the shells of crustaceans.

Clade. A grouping of organisms that includes a single common ancestor and all of its descendants. Example being Dinosauria is the clade that includes all dinosaurs.

Clay. A fine-grained geological material that develops plasticity when wet and becomes hard and brittle when dry. Grain size of less than 0.002 mm.

Clasts. A fragment of rock, ranging from less than 1 μm to as a big as an apartment block.

Cliff. A mass of rock, usually a rock face, which rises high and is almost vertical and maybe overhanging.

Concretion. A rounded mass of minerals that forms within a sedimentary rock.

Conglomerate. A coarse-grained sedimentary rock composed of other rock fragments in a cementing matrix.

Continental shelf. The portion of the continent that is submerged under shallow water, also known as a shelf sea.

Coprolite. Fossilised faeces.

Coarse grained. A rock or sediment where the grains are visible to the naked eye.

Crevasse splay. A fluvial deposit when a stream or river bursts its banks and deposits sediments onto the floodplain.

Cretaceous. A geological period in the Mesozoic Era extending from 145 – 66 million years before present.

Crocodile. A large predatory semiaquatic reptile with long and narrow jaw compared to the wider and short snouted alligator. Known for saying 'in a while'.

Cross bedding. Layering within the strata where layers intersect at an angle to the main bedding, usually representing ripple marks.

Crystallography. A branch of science that studies a crystals atomic and geometric structure of the crystal lattice.

D

Decomposing/ Decomposition. The state or process of rotting or decaying.

Desiccation cracks. Also known as mud cracks. Sedimentary structures formed as water evaporates from a clay rich sediment, shrinking the clays, resulting in cracking.

Deposition. The geological process where sediments, soil and rocks are transported and added to a landform or landmass. Wind, flowing water, the sea, ice or gravity can transport weathered material, and it is deposited when the forces responsible for sediment transportation are no longer sufficient to move the material.

Depositional environment. The environment associated with the physical, chemical and biological processes associated with the deposition of a particular type of sediment, and rock types that form, mostly based on modern environments.

Dinosaur. A diverse group of reptiles of the clade Dinosauria, inhabiting the Earth from 245 million years ago until the present day, as birds of course, are dinosaurs!

Durophagous. Animals that consume hard-shelled prey, such as molluscs or crabs.

E

Embayment. The coastal recess creating a bay.

Environment. The surroundings or conditions in which a person, animal, or plant lives or operates.

Eocene. A geological epoch in the Paleogene period, extending from 56– 33.9 million years before present.

Epifauna. Animals that live on or are attached to the seafloor.

Estuarine. An area where a freshwater river or stream meets the sea. The salt and freshwater mix, resulting in brackish water, which is slightly salty, but not as salty as seawater.

Ephemeral lake. A lake that is usually dry but fills with water for brief periods during or after precipitation.

Erosion. The process in which rocks and sediments are removed, from actions such as water flow or wind, or dissolution, and then transported to another location where it is deposited.

Erosional surface. The surface of a rock or bed that was formed by erosion.

Evaporite. Water soluble sedimentary material / mineral, which is produced by evaporation of a concentrated aqueous solution.

Extant. Animals that still exist.

Extinct. Animals that no longer exist, having completely died out.

Exposure. An area of rock that is visible either naturally or artificially.

F

Fauna. All animal life present in a particular area or time.

Fault. A fracture of zone of fractures between two units of rock. The fault allows for the blocks to move relative to each other, occurring rapidly or over an extended period of time.

Fine grained. A sediment where the grains are too small to see without magnification.

Fines upwards. The grainsize decreases as you go up a bed. This could be grit at the bottom and a fine sand or clay at the top. This indicates a decrease in energy in the depositional environment, such as a slowing down of a river's flow.

Flaser bedding. A sedimentary structure with alternating layers of rippled sand and mud.

Flints. A sedimentary rock consisting of microscopic, nearly undetectable cryptocrystalline quartz crystals.

Flora. All plant life present in a particular area or time.

Fluvial. Sediments that have been deposited by a stream or river.

Fluviatile. Relating to rivers or found in/near rivers.

Flint tools. Tools carved from flint. Dating from 3300 B.C. to 3 million years.

Freshwater. Water with no salt content.

Formation. A body of rock having a consistent set of physical characteristics (lithology) that distinguishes it from adjacent bodies of rock, and which occupies a particular position in the layers of rock exposed in a geographical region.

Fossil. The preserved remains or trace remains of ancient organisms.

Fossiliferous. A unit of rock or strata bearing or containing fossils.

G

Gastrolith (Stomach stone). A rock held within the gastrointestinal tract of an animal. These are used to help grind tough plant materials or used as a ballast in aquatic/semiaquatic animals.

Gastropod. A mollusc within the class Gastropoda, such as snails or slugs.

Geological structure. The result of powerful tectonic forces that occur within the earth.

Geology. The study of the Earth's physical structure, composition and history.

Geologist. A scientist who studies geology.

Glacial. In reference to processes and features related to glaciers. Or to describe the presence of cold, icy conditions during Ice Age events.

Glauconitic. A rock/clay or sandstone containing glauconite grains to make the colour of the rock green.

Group. A lithostratigraphic unit in the rock record, divided into individual formations.

Grande Coupure. An important faunal turnover event in early Oligocene Europe.

H

Habitat. The natural home or environment of an animal, plant, or other organism.

Hauterivian. A geological stage in the Lower Cretaceous, extending from 132.6 – 125.7 million years before present.

Heterolithic. A sedimentary structure consisting of alternating layers of sand and mud, formed mainly in tidal flats, delta fronts and floodplains.

Horizon. A specific layer within the rock record, usually a bedding surface where there is a marked change in lithology.

Holocene. A geological epoch in the Quaternary, lasting 11,700 years ago to the present day.

Hypersaline. An aquatic environment that has saline levels surpassing that of ocean water.

I

Indigitates. Folds, boundaries or shorelines that have projections extending like fingers locking together.

In situ. The original position of a rock or fossil within the surrounding geology.

Interbedded. When beds of different lithologies lie between or alternate with beds of a different lithology.

Interglacial. A geological period of a warmer average global temperature, lasting thousands of years, separating glacial periods within an Ice Age.

Invertebrate. An animal that does not have a vertebral column (backbone), such as snails, worms or crustaceans.

J

Jurassic. A geological period in the Mesozoic Era extending from 201.4 – 145 million years before present.

K

K-Pg boundary. Also known as the **K-T boundary**. The boundary marking the end of the Cretaceous Period and the last of the Mesozoic Era, around 66 million years ago. Globally this is an ash layer horizon containing a high percentage of iridium, in association with a meteorite impact This has been associated with a mass extinction event, destroying a majority of the world's Mesozoic species.

Konzentrat-Lagerstätten (Concentration Lagerstätten). A deposit with a high concentration of fossils.

Konservat- Lagerstätten. (Conservation Lagerstätten). A deposit known for the extraordinary preservation of fossils, where soft parts are preserved.

L

Lacustrine. Relating to or associated with lakes.

Lagerstätte (plural Lagerstätten). A deposit with unusual occurrences of relatively well-preserved fossils.

Lagoonal. Relating to a shallow body of water separated from larger water bodies by small land barriers.

Lateral variation. The changes in geology in a set layer across the length of the exposure.

Lenticular bedding. A sedimentary bedding patterning of alternating layers of mud and sand. Formed when mud falls out of suspension during periods of slack water (no or little water movement) covering small formations of sand.

Life position. A fossil in life position is where it has been preserved in the same orientation as it was when it lived, such as a tree stump still upright.

Lignite. A brownish black coal, in which the texture of the original wood is distinct.

Limestone. A hard sedimentary rock composed mainly of calcium carbonate or dolomite, such as chalk.

Lithology. Description of the physical characteristics of a rock unit, such as colour, texture and grain size.

Lithostratigraphic unit. A body of rock that is defined and identified by its observable physical characteristics or lithology.

Lustre. The surface appearance of a mineral and how it reflects light.

Lutetian. A stage within the Eocene, spanning 48 – 41 million years before present.

M

Maastrichtian. A geological age in the Late Cretaceous epoch, dating to 72.2 – 66 million years.

Marine. Relating to or found in the sea.

Marginal marine. An environment between the land and the sea, such as lagoons and estuaries.

Marine transgression. A geological event during which the sea level rises relative to land, resulting in flooding and shorelines moving to higher ground.

Marine regression. A geological event during which the sea level falls relative to land, resulting in areas of submerged seafloor to be exposed above sea level.

Marine reptile. A reptile that has become secondarily adapted for an aquatic lifestyle.

Marls. A soft, light-coloured sedimentary rock made of clay and calcium carbonate.

Matrix. A groundmass of fine-grained material in which larger grains/fossils are embedded.

Megalith. A large stone that forms part or all of a prehistoric monument.

Member. A Lithologically distinct part of a formation.

Mesolithic. Or Middle Stone Age. Archaeological period dating to 11,700 – 6,000 years.

Mica. A shiny silicate mineral with a layered structure.

Microfossil. A fossil that can only be seen with a microscope.

Mineral. A naturally occurring inorganic solid, with a definite chemical composition and an ordered atomic arrangement.

Mineral staining. When groundwater full of minerals seeps through a rock. The minerals are left behind and crystallise, occasionally causing the rock colour to change.

Miocene. The first epoch of the Neogene, extending 23 – 5.3 million years ago before present.

Mollusc. An invertebrate animal with a soft body, no spine and often covered with a shell. Such as snails, slugs, clams and squid.

Monocline. A step-like fold in the rock layers that has one side steeply dipping, and one side horizontal or gently dipping.

Mottled. A layer marked by a patchy or blotchy pattern of colours or shades.

Mudstone. A dark sedimentary rock formed from consolidated mud, lacking lamination.

N

Neogene. A geological period in the Cenozoic, extending from the Paleogene Period to the Quaternary, 23.03 – 2.58 million years before present.

Neolithic. Or New Stone Age. Archaeological period dating to 6,000 – 2,000 years.

Neomorphing. A replacement of one mineral by a different crystal form of the same mineral, such as aragonite into calcite.

Neritic. The shallow part of the sea near a coast.

Niche. The role an organism plays in a community or ecosystem.

O

Offshore environment. The areas beneath the surface of a large body of water (seabed).

Oligocene. A geological epoch in the Paleogene period, extending from 33.9 – 23 million years before present.

Omission surface. Stratigraphic surfaces swept bare by erosion or starvation of sediment, and which have usually been modified by benthic organisms such as encrusting molluscs (oysters, bryozoa) or borings.

Omomyidae. A family of small tarsier-like primates that were common during the Eocene of Europe and North America.

Ophiomorpha. An ichnotaxon or trace fossil, characterised by vertical or steeply inclined burrows. These are interpreted as a crustacean burrow living in a near shore environment.

Ornithopod/ ornithischian. A bipedal herbivorous dinosaur.

Orogeny. The primary mechanism by which mountains are formed due to a collision between two landmasses.

Outcrop. A rock formation that is visible on the surface.

Overbank. Sediment that has been deposited on the floodplain of a river or stream that have variable rates of water discharge.

Oxidation. A chemical weathering process that occurs when oxygen and water react with chemical compounds, usually iron-rich minerals, creating rusty brown oxides, and eventually breaking down a rock.

P

Palaeobiology. The study of the biology of extinct animals and plants.

Palaeolithic. Or Old Stone Age. Archaeological period dating to 900,000 – 11,700 years.

Palaeothere. The palaeotheriidae. A family of perissodactyls closely related but not ancestral to horses that were widespread in Eocene Europe. It includes genera such as Palaeotherium and Plagiolophus.

Paleocene. A geological epoch in the Paleogene period, extending from 66 – 56 million years before present.

Paleoenvironment. An environment that has been preserved in the rock record.

Paleoflora. The collection of plant fossils that existed during a specific geological period.

Paleosol. 'Fossil soil' A former soil preserved by burial underneath sediments, in which the soil has turned into rock/clay.

Paleogene. Geological period in the Cenozoic, spanning from the end of the Cretaceous to the Neogene, extending from 66 – 23.03 million years before present.

Palaeontology. The scientific study of ancient life dealing with fossils.

Palaeontologist. A scientist who studies fossils and aspects of ancient life.

Paleovalley. A buried or ancient river valley that is no longer active, often filled with sediment from subsequent geological processes.

Patina. A gloss or sheen on a surface resulting from age or polishing.

Pedogenic clays. Preserved ancient soils which have turned into clay. See Paleosol.

Periglacial. Environment with a cold climate, with permafrost and seasonal thawing.

Perissodactyl. The Perissodactyla. An order of ungulates known as the 'odd-toed ungulates' that includes horses, tapirs, rhinoceroses and their extinct relatives.

Permafrost. Where the ground freezes and remains frozen for two consecutive years or more, at or below 0°C.

Phylum. A major taxonomic rank, below Kingdom, grouping organisms that are related by the same evolutionary histories, example being Mollusca (snails, clams, squid), Arthropoda (insects, spiders, crustaceans) and Chordata (animals with a notochord, such as mammals, birds, fish, reptiles).

Physical process. Natural processes and forces which shape the physical planet.

Phytoplankton. Microscopic marine algae that float in the upper part of the ocean's waters, carried by currents and tides. They contain chlorophyll and require sunlight to live and grow.

Planispiral. Coiled in a flat, single plane forming a disk like or flattened spiral shape.

Plankton. A microscopic organism that lives in ocean waters, carried by tides and currents.

Plant debris. Woody and plant-based debris accumulated in a bed

Pleistocene. A geological epoch in the Quaternary, lasting 2,580,000 to 11,700 years ago.

Plesiosaur. A group of extinct long necked marine reptiles ranging from the Upper Triassic to the Upper Cretaceous.

Pseudomorph. A crystal that has been replaced by another substance, but still retains it's original shape.

Pterosaur. A group of extinct flying reptiles that lived between the Upper Triassic and Upper Cretaceous.

Pyritised. The replacement of the original material by iron pyrite.

Q

Quaternary. The most recent period in the Cenozoic era, spanning from 2.58 million years ago to the present.

R

Radiation (adaptive radiation). A rapid diversification of a single ancestral species into many new species, usually seen after a mass extinction.

Reduction. A process where there's a loss of oxygen or a gain of hydrogen, causing bleaching features in red bed sediments.

Regression. A sea level fall, exposing areas of submerged seafloor.

Reworked fossil. Fossils that have been moved from their original location and deposited into younger sediments.

Rootlets. A small root or division of a root.

S

Salinity. The concentration of dissolved salts in water.

Saline. Water containing a significant amount of dissolved salt.

Sandstone. A sedimentary rock made of sand grains, ranging from 0.05 - 2 mm in diameter.

Sauropod. A group of quadrupedal dinosaurs with long necks and tails.

Sediment. A solid material, consisting of rocks, minerals and remains of plants and animals, which has been broken down by weathering or erosion and transported to another area by physical processes (wind, water etc.).

Sedimentary rock. A rock composed of smaller rocks or fossils.

Sessile. An organism that lacks the ability to move and predominantly immobile or are attached to the substrate.

Shelf. Also known as the 'continental shelf'. The portion of the continent that is submerged under shallow water, from the shoreline to 100-200 m water depth.

Shoreface. The area seaward from the low tide shoreline to where the wave action on the seabed becomes minor.

Sideritic nodules. A heavy, brown or black concretion made of the mineral siderite (a iron carbonate mineral, FeCO3).

Slit. A fine-grained geological material that becomes slippery or soapy when wet and floury when dry. Grain size of 0.002 – 0.05 mm.

Solifluction. When water saturated soil moves down a slope due to freeze-thaw cycles.

SSSI (site of special scientific interest). A formal conservation designation describing an area that is of interest to science, such as having rare species of plant or animal or important geological features.

Spring. A natural exit point at which groundwater emerges out of the aquifer and flows across the surface.

Stratigraphy. The study of the arrangement and succession of rock layers (strata) and layering (stratification). As well as the origin, composition and distribution of the geological strata.

Steinkern. The infilling of a fossil hollow, consisting of mud of sediment that has hardened to produce an internal mould.

Subaerial dissolution. A type of chemical weathering where the dissolved minerals in rainwater (or ground water) react with and breaks down the rocks on the land surface.

Syncline. A downward fold of stratified rock, creating a concave 'U' shape.

T

Taxon. A general term for any named group of organisms at any rank in the classification system.

Taxa. Plural of taxon.

Thanetian. The uppermost stage of the Paleocene Epoch, spanning 59 – 56 million years ago.

Theropod. Bipedal carnivorous dinosaurs.

Trace fossil. A fossil that provides indirect evidence of ancient plants or animals, such as footprints, nests, burrows or faeces.

Transgression. A sea level rise, with sea water flooding and submerging previously exposed land.

Triassic. The earliest period of the Mesozoic, spanning 251.9 – 201.4 million years before present day.

U

Unconformity. A boundary between rock layers where there has been a break in deposition of sediment or when erosion occurs.

V

Vertebral series. The row of vertebrae going from the head to the end of the tail.

Vertebrate. Animals that have a vertebral column (backbone) and an internal skeleton.

Y

Ypresian. The oldest stage in the Eocene, spanning 56 – 48 million years before present.

Z

Zooplankton. A microscopic organism that lives in ocean waters, carried by tides and currents. These feed on other plants and animals.

Zygopophysis. The articular process of the neural arch of a vertebrae.

Index

Fossil Notes

Date

Fossil Notes

Date